EU ENLARGEMENT, REGION BUILDING AND SHIFTING BORDERS OF INCLUSION AND EXCLUSION

BORDER REGIONS SERIES

Series Editor: Doris Wastl-Walter

In recent years, borders have taken on an immense significance. Throughout the world they have shifted, been constructed and dismantled, and become physical barriers between socio-political ideologies. They may separate societies with very different cultures, histories, national identities or economic power, or divide people of the same ethnic or cultural identity.

As manifestations of some of the world's key political, economic, societal and cultural issues, borders and border regions have received much academic attention over the past decade. This valuable series publishes high quality research monographs and edited comparative volumes that deal with all aspects of border regions, both empirically and theoretically. It will appeal to scholars interested in border regions and geopolitical issues across the whole range of social sciences

EU Enlargement, Region Building and Shifting Borders of Inclusion and Exclusion

JAMES WESLEY SCOTT

Leibniz-Institute for Regional Development and Structural Planning, Germany

ASHGATE

Published by
Ashgate Publishing Limited
Gower House
Croft Road
Aldershot
Hampshire GU11 3HR
England

Ashgate Publishing Company
Suite 420
101 Cherry Street
Burlington, VT 05401-4405
USA

Ashgate website: http://www.ashgate.com

British Library Cataloguing in Publication Data
EU enlargement, region building and shifting borders of
    inclusion and exclusion. – (Border regions series)
    1. European Union – Membership 2. International cooperation
    3. European Union countries – Boundaries 4. Europe – Economic
    integration 5. European Union countries – Relations
    I. Scott, James Wesley
    337.1'42

Library of Congress Control Number: 2006922519

ISBN-10: 0 7546 4542 8
ISBN-13: 978-0-7546-4542-9

Printed and bound in Great Britain by TJ International Ltd, Padstow, Cornwall.

Table of Contents

List of Figures   *vii*
List of Maps   *ix*
List of Tables   *xi*
List of Abbreviations   *xiii*

Part I   Introduction: Aims and Outline of the Book   **1**

1   Wider Europe as a Backdrop   3
    James Wesley Scott

Part II   Borders and the Geopolitics of EU Enlargement   **15**

2   Wider Europe: Geopolitics of Inclusion and Exclusion at the EU's
    New External Boundaries   17
    James Wesley Scott

3   Geopolitics of Scale and Cross-Border Cooperation in Eastern
    Europe: The Case of the Romanian-Ukrainian-Moldovan Borderlands   35
    Gabriel Popescu

4   The European Community as a Gated Community: Between Security
    and Selective Access   53
    Henk van Houtum and Roos Pijpers

Part III EU Enlargement and Its Impact at New External Borders   **63**

5   Changing Border Situations within the Context of Hungarian Geopolitics   65
    Zoltán Hajdú and Imre Nagy

6   The Impact of EU Enlargement on the External and Internal Borders
    of the New Neighbours: The Case of Ukraine   81
    Olga Mrinska

7   Regional Cooperation in the Ukrainian-Russian Borderlands:
    Wider Europe or Post-Soviet Integration?   95
    Tatiana Zhurzhenko

8   The New Neighbourhood – A 'Constitution' for Cross-Border
     Cooperation?   113
     Ilkka Liikanen and Petri Virtanen

**Part IV Evolving Cooperation Frameworks and Cross-Border
Regional Development**   **131**

9   The Impact of EU Enlargement on Moldovan-Romanian Relationships   133
     Alla Skvortova

10   Euroregions along the Eastern Borders of Hungary: A Question of Scale? 149
     Béla Baranyi

11   Transboundary Interaction in the Hungarian-Romanian Border Region:
     A Local View   163
     Gyula Szabó and Gábor Koncz

12   Patterns of Legal and Illegal Employment of Foreigners along the
     Hungarian-Ukrainian Border   171
     István Balscók and László Dancs

13   Local and Regional Cross-Border Cooperation between Poland
     and Ukraine   177
     Katarzyna Krok and Maciej Smętkowski

**Part V   Cross-Border Cooperation and Regional Development at the
Former External Borders**   **193**

14   Normalizing Polish-German Relations: Cross-Border Cooperation
     in Regional Development   195
     Grzegorz Gorzelak

15   Regional Development in Times of Economic Crisis and Population
     Loss: The Case of Germany's Eastern Border Regionalism   207
     Hans-Joachim Bürkner

Bibliography   *217*
Index   *239*

List of Figures

6.1   Relative macroeconomic indicators of Ukraine and some
neighbouring countries in central and eastern Europe   92

12.1   The development of rail and road traffic at crossings on the
Hungarian-Ukrainian border, 1988–2002   173

12.2   The distribution of foreigners with valid work permits by
counties (31 December 1999)   174

13.1   Polish-Ukrainian border traffic,1992–2003   183

13.2   Value of Polish-Ukrainian foreign trade, 1992–2003 (in million US$)   185

13.3   Impact of Poland's EU accession on the conditions of trade exchange
with Ukraine as seen by Polish entrepreneurs   189

List of Maps

3.1   Euroregions between Romania, Ukraine and Moldova   39

5.1   Hungary and its neighbours   67

6.1   Industrial growth in Ukrainian regions, 2001   86

6.2   Industrial growth in Ukrainian regions, 2003   87

8.1   Euregio Karelia   125

10.1 Euroregions and regional cross-border cooperation in eastern
      Hungary, 2002   152

10.2 Axes of economic development in the territory of the DKMT
      Euroregion   153

10.3 Potential and developing border regions and interregional
      cooperation initiatives   156

10.4 The Hajdu-Bihar-Bihor (county-level) and Bihar-Bihor
      (microregional) Euroregions   158

11.1 Settlements studied during the survey   164

12.1 Communities studied during the survey   175

13.1 The Polish-Ukrainian border regions   181

14.1 Polish territories lost to Prussia and Germany, 1803–1945   197

14.2 Polish territories eligible for Poland-Germany Phare CBC grants   200

14.3 The German-Polish border region and Euroregions   203

15.1 The German-Polish Euroregion Pro-Europa Viadrina   208

List of Tables

6.1   Ukrainian regions and their relative economic weight, 2003   84

9.1   Trade between Romania and Moldova, 1998–2004 (in million US$)   139

9.2   Implemented and on-going TACIS projects directly contributing to the development of Moldovan-Romanian cooperation   144

11.1   Frequency of cross-border contacts (percentage of answers)   166

11.2   What is the name of the neighbouring settlement in Romania? (percentage of answers)   167

11.3   How can your settlement contribute to the establishment of cross-border cooperation with a neighbouring settlement? (percentage of first answers received)   168

11.4   How can your settlement contribute to the establishment of cross-border cooperation with a neighbouring settlement? (most frequent answers received)   169

13.1   Social and territorial structure of the border region, 2001   180

13.2   Trade turnover of Polish border voivodships with Ukraine   186

List of Abbreviations

AMU	Arab Maghreb Union
APEC	Asia Pacific Economic Cooperation
BSEC	Black Sea Economic Cooperation
BSR	Baltic Sea Regionalism
CARDS	Community Assistance for Reconstruction, Development and Stabilisation
CBC	cross-border cooperation
CEFTA	Central European Free Trade Area
CFSP	Common Foreign and Security Policy (of the European Union)
CIS	Commonwealth of Independent States
COMECON	Council for Mutual Economic Assistance
DKMT	Danube-Körös-Maros-Tisza Euroregion
EBRD	European Bank for Reconstruction and Development
ECHR	European Court of Human Rights
EEC	European Economic Community
EFTA	European Free Trade Area
EIB	European Investment Bank
EMP	Euro-Mediterranean Partnership
ENP	European Neighbourhood Policy
ENPI	European Neighbourhood and Partnership Instrument
ESC	Economic and Social Committee (of the European Union)
ESPD	European Spatial Development Policy
ESPON	European Spatial Planning Observation Network
EU	European Union
EvrAzES	Evraziiskoe Ekonomicheskoe Soobshchestvo (Eurasian Economic Community)
FDI	foreign direct investment
FTZ	Free Trade Zone
GDR	German Democratic Republic
GUUAM	Georgia, Ukraine, Uzbekistan, Azerbaijan and Moldova
IFI	International Financial Institutions
INCD Delta Dunarii	Institutul National de Cercetare – Dezvoltare
INTERREG	(for Interregional) Community Initiative for the ERDF (European Regional Development Fund)
ISPA	Instrument for Structural Policies for Pre-Accession

MEDA	Mediterranean European Development Assistance
MERCOSUR	Mercado Común del Sur
MNC	Mediterranean non-EU member countries
MSzMP	Magyar Szocialista Munkáspárt (Hungarian Socialist Workers' Party)
MSZP	Magyar Szocialista Párt (Hungarian Socialist Party)
NAFTA	North American Free Trade Area
NATO	North Atlantic Treaty Organization
ND	Northern Dimensional
NDEP	Northern Dimension Environmental Partnership
NEFCO	Nordic Environment Finance Corporation
NGO	Non-governmental organization
NIB	Nordic Investment Bank
NIS	New Independent States
NNI	New Neighbourhood Initiative
NNP	New Neighbourhood Policy
NOPEF	Nordic Project Fund
OTKA	Országos Tudományos Kutatási Alapprogramokról (Hungarian National Research Fund)
PCA	Partnership and Cooperation Agreement
PHARE	Poland/Hungary Aid for the Reconstruction of the Economy
PHARE-CREDO	EU-Phare Credo programme – Cross-border cooperation between Central European countries
PHARE-SPF	PHARE Small Projects Fund
PKP	Polskie Koleje Państwowe (Polish National Railways)
REGON	Rejestr podmiotów gospodarki narodowej (Polish national economic registry)
SAP	Stabilisation and Association Process
SAPARD	Special Accession Programme for Agriculture and Rural Development
SECI	South-Eastern Cooperation Initiative
SME	small and medium-sized enterprises
SP	Stability Pact
TACIS	Technical Assistance to the Commonwealth of Independent States
TNC	Transnational Corporation
WTO	World Trade Organization

PART I
Introduction:
Aims and Outlines of the Book

Chapter 1

Wider Europe as a Backdrop

James Wesley Scott

2004 marked an historic year in the process of European enlargement; ten Central and Eastern European states, as well as Malta and Cyprus, joined the European Union. A new round is likely in 2007 as Romania and Bulgaria prepare for membership. Above and beyond the future EU-27, debate as to the perspectives of future enlargement, and the status of relationships between the EU and 'new' neighbours such as Turkey and the Ukraine is intensifying. As the EU takes on new members and its external boundaries gradually shift, socioeconomic and political transformations are taking place at the borders that not only portend new regional development opportunities but also many potential problems and tensions. With Wider Europe and its long-term commitment to support local and regional initiatives of cross-border cooperation, the EU has expressed a will to avoid future divisions between 'East' and 'West' and 'North' and 'South'. This is to be achieved through comprehensive cooperation agendas that transcend political, economic and cultural dividing lines and that address socioeconomic disparities, political tensions and potential conflicts of interest. In addition, a New Neighbourhood Instrument (NNI) is in the making that – theoretically – will improve the material and institutional conditions for cooperation with 'non-EU Europe'.

And yet, it is unclear to what extent such regional partnerships between the EU and non-EU states can move beyond market logics and embrace heterogeneous economic and sociopolitical realities. The EU insists that enlargement will not signify 'new divisions' in Europe, but the processes of inclusion and exclusion and imposition of visa restrictions on non-EU citizens could pose new obstacles to cooperation, conjuring fears of an emerging 'fortress Europe' that effectively divides the continent. While some elements of EU policy work to enhance cooperation and cross-border interaction, others work against this. An agenda aiming at securing borders and implementing selective border regimes as well as more basic economic protectionism are the obverse of cooperation incentives and diplomatic attempts at 'inclusion' (Scott 2005).

Ultimately, these questions of interstate relationships and political community are about borders and their wider significance. At the new (and future) external borders of the EU it will be necessary to find mechanisms that mediate between external pressures and local concerns and transcend socioeconomic, political and systemic asymmetries. Clearly, however, our academic and everyday perceptions of borders, border regions and cross-border cooperation continue to be affected by overlying geopolitical events, reflecting the concerns of the times and processes of EU integration and enlargement. European research has also focused on the significance of borders and cross-border cooperation within processes of European integration and 're-scaling'.[1] In fact, RTD projects funded by the EU (for example within the scope of the Fifth Framework Programme) have highlighted the multilevel significance of borders in organizing social, political, economic and cultural life.[2] The European research projects 'EXLINEA' and 'EU BORDER IDENTITIES', for example, indicate that border regions are often characterized by a strong local identity based on 'we'/'them' categories at the same time that Europe, in the guise of EU programmes, Euroregions (as an instrument of governance and rapprochement), etc., tends to be more visible at the border. The question as to whether border regions can function as laboratories of cooperation and/or postnational political community remains an important one that should not be neglected. This research indicates, furthermore, that cross-border cooperation is a very selective project of networking and 'region-building'. Given the simultaneity of inclusion and exclusion dynamics and discourses that characterize many borderlands contexts, the quality of cooperation will, to a great extent, depend on the role regional stakeholders and/or political elites assume in promoting a regional idea and bridging political/cultural differences. The quality of political messages of cross-border cooperation, however, is not only a local issue; it is subject to practices and discourses that operate at several different spatial levels and societal realms.

Multilevel Research Perspectives

These research perspectives have contributed to the fact that borders are now largely understood to be multifaceted social institutions rather than mere markers of state sovereignty. Presently, there is no single theory, concept or discourse on borders that enjoys predominance within the European context (Newman 2003). Generally speaking, much current research on European borders is characterized by problem-oriented but socially critical readings of 'bordering' processes within the context of European integration and enlargement. Practical issues of cross-border cooperation are thus interpreted not only in terms of 'technical' issues and 'structural' constraints but increasingly in terms of the perception of borders as symbols and identity-constructing elements as well. Liam O'Dowd (2002, p. 29), for example, argues that 'one of the key lessons to be drawn from the history of state formation in Europe is that the structure, functions and meanings of state borders seldom remain fixed or stable for long periods.' Within this historical context of change O'Dowd sees the European project as reconfiguring borders in terms of both 'barriers' and 'bridges'.

---

1   See, for example, Kramsch (2003 and 2004).

2   See the websites of the following FP 5 projects: EUBORDERCONF <www.euborderconf.bham.ac.uk>, EU BORDER IDENTITIES <www.borderidentities.com> and EXLINEA <www.exlinea.org)>.

However, O'Dowd (2002, p. 30) also acknowledges the multilevel contingency of cross-border interaction; heterogeneity is the rule and generalizations about cross-border practices are often difficult to justify: 'Heterogeneity arises from different experiences of border formation, and formal and informal cross-border relationships, along with the relative economic and political power of contiguous states and the role, if any, played by external powers or regional ethnic and national questions. Moreover, the EU's stress on market integration and economic competitiveness impacts in differential ways on pre-existing border heterogeneity'.

This multilevel approach is reflected in current research on cross-border cooperation in Europe. Within the EXLINEA context, discourses and practices concerning cross-border cooperation are seen as constituting 'regionalization' processes. In this case 'regionalization' refers to the development of institutions and social practices that help create a framework for the solution of common problems and the resolution of conflicts in different regions at the EU's external boundaries. The essays in the present volume are based on the EXLINEA research project, funded by the European Union and the Fifth Framework Programme (contract HPSE-CT-2002-00141; <www.exlinea.org>). By focusing on discourses and practices, EXLINEA has attempted to illustrate how, at different phases of EU enlargement, cross-border cooperation has been and continues to be conditioned by overlying geopolitical considerations and national/local development contexts.

Given the timeliness of the individual contributions, the principal thrust of the book will lie in case studies where actual parameters and experiences of cross-border governance will be analyzed. At the same time, the essays contained in this book are characterized by a theoretically informed focus on practical issues of cross-border cooperation. This manifests itself in a discussion of 'bordering' and 'rebordering' processes within and without the EU-25 and in an examination of capacities for 'region-building' across national borders in central and eastern Europe. The cooperation initiatives under scrutiny involve, among others, cooperative structures, governance practices, conflict-minimizing dialogue and strategies for joint economic development. Some of the questions addressed by this book include the following:

- What are the principle political and socioeconomic challenges that require cross-border collective action in contexts of EU enlargement?
- How are policies that regulate the 'permeability' of borders compatible with pursuits to promote cross-border cooperation? How do European and national political policies and interests coalesce and/or clash with regard to the development of closer cross-border cooperation?
- How is transnational political community being (re)defined at the EU's external borders?
- What have been the local responses to EU border policies at former and present external boundaries of the EU?
- What strategies have been developed by local actors to promote region-building in terms of formal and informal integration processes? What governance functions can be realistically attributed to cross-border cooperation?

- What results in terms of perceived added value have been achieved through cross-border cooperation?
- Is cross-border region-building helping instil a sense of common interest and European identity?

How this Book is Structured

The book is organized along four thematic sections. The first part deals with bordering and geopolitics of EU-Enlargement. In his opening essay, James Scott discusses the development of the 'Wider Europe' strategy (WE) as an overlying political context for cross-border cooperation. As the political will to proceed with further enlargements recedes, an alternative strategy of 'inclusion' of neighbouring states is seen as imperative in order to avoid future divisions between 'East' and 'West' and 'North' and 'South'. Partnership and free trade are the integrative mechanisms that will substitute for direct membership in the EU. Furthermore, the EU sees its WE strategy as transcending the limited focus of other geopolitical visions (such as NAFTA) by promoting comprehensive cooperation agendas that include not only economic but also political and cultural issues. In addition, a New Neighbourhood Instrument (NNI) is in the making that – theoretically – will improve the material and institutional conditions for cooperation with the European states outside the EU and Mediterranean states. Scott therefore focuses on the possible consequences of WE for cross-border cooperation at the new external borders of the EU. Among other countries, border communities in Finland, Estonia, Poland, Slovakia and Hungary have an interest in maintaining well-functioning ties with neighbouring localities and regions in the former Soviet Union. However, it is unclear to what extent such regional partnerships between the EU and 'non-EU' states can move beyond market logics and 'embrace' heterogeneous economic and sociopolitical realities.

Gabriel Popescu continues the discussion of geopolitical 'bordering' with a treatment of Romanian-Moldavian relations and cross-border cooperation. In his paper Popescu sets out to demonstrate how multiscalar national, supra- and subnational forces interact to shape social life in Eastern European borderlands within the context of European integration. In one way the European Union envisions integration through institutionalized cross-border cooperation in the form of Euroregions and based on the principle of subsidiarity; local authorities are seen to be better prepared than central governments to address the needs of local inhabitants. With Euroregions, a cooperation instrument developed in 'core' EU-Europe, the EU pursues a transnational integration strategy which aims at redefining (and rescaling) Westphalian state territoriality in order to better address the challenges of a 'world of flows'. However, in eastern Europe, an area of complex social, economic and political diversity, Euroregions have been partially 'co-opted' by East European governments in order to benefit from EU support. As a result, Popescu argues, Euroregions, such as those emerging in the Romanian-Moldavian case, are 'top-down' creations and a real devolution of power to local authorities has not taken place. In the case of Romania and Moldavia the state maintains complete control over the decision process in Euroregions, reducing them to mere political tools in international politics.

Roos Pijpers and Henk van Houtum take this geopolitical discussion beyond physical boundaries. They employ the metaphor of a 'gated community' to characterize, as they see it, policies of selective access to the European Union. Pijpers and van Houtum argue that fear of mass (in)migration remains an important aspect of ongoing processes of sociospatial bordering in and by the European Union (EU) and its various member states. The 2004 enlargement processes have intensified these fears and, as a result, increased ambivalence towards the project of European integration. Reasons for this are evident: the EU-25 is characterized by particularly sharp socioeconomic disparities and greatly varying income levels. Cultural and linguistic barriers within the EU have also become much more complex. Furthermore, with the EU's geographical shift eastward, the external borders of the Union have become symbolic markers of difference, both in terms of social welfare standards and stages of systemic transformation. As has always been the case, the internal liberalization of capital and labour markets and guarantees of 'moral equality' for EU citizens go hand in hand with a tightening of management and control efforts at the Union's external borders. At the same time attempts to strategically select 'desirable' immigrants from outside the EU are increasing in order to bypass labour shortages in specific segments of the economy. Pijpers and van Houtum thus contribute to a critical understanding of the normative justifications given for these at once 'exclusionary' and 'strategically selective' bordering processes in the EU. The argument will be made that the moral panic regarding immigration and possible future migration policies in the majority of EU member states follows a strategic logic that much resembles the management of a 'gated community'.

In the second part of this book, EU enlargement and its impacts at new external borders will be scrutinized. Imre Nagy and Zoltan Hajdú begin the section with a perusal of the changing significance of Hungary's borders as a result of the new political situation in Europe. Since 1989 Hungary's geopolitical situation has been fundamentally transformed. With the end of bipolar systemic confrontation, Hungary – as well as its postsocialist neighbours – has been seeking to redefine its international relations and its own sense of purpose with an enlarging European Union. Systemic and economic transformation have resulted in a fundamental realignment of political orientations, allegiances and, among other things, membership in the EU and NATO. At the same time Hungary has also been able to develop a new quality of bilateral relationships with its immediate neighbours. As the authors argue, the significance of borders in Hungary is therefore pervasive and affects the entire national territory. Reasons for this are historical and include the existence of sizeable Hungarian-speaking minorities within border regions of neighbouring states. Hence, within the post–cold war context, the relative importance of Hungary's external borders in terms of cross-border interaction has greatly increased. This applies both to old neighbours (Austria and Romania) and states that have recently become independent (Slovakia, the Ukraine, Serbia, Croatia and Slovenia). This essay will systematically characterize the shifting geopolitical situations of Hungary's various borders based on a variety of criteria. The primary objective will be to differentiate between 'internal' and 'external' borders based not only on membership in international organizations (such as the EU and NATO) but also based on national and regional interests and motivations that inform cross-border interaction.

Olga Mrinska takes our discussion to the Ukraine, where the 'arrival' of the EU as a neighbour will have far-reaching consequences for the country's spatial development. Historically, Ukraine has been divided by what could be provocatively called a 'Huntington line' into two major parts: a 'West' oriented towards central and western Europe and an 'East' oriented towards Russia. Western Ukraine's common history with Poland, Slovakia, Hungary and Romania has promoted deeper ties with the European community. This is a matter not only of formal relations but also of informal relations borne out of social and cultural similarities. The history of eastern Ukraine, on the other hand, has for centuries been closely tied to that of Russia; this part of the Ukraine remains heavily influenced by Russian culture and the Russian language. Hence, the possible emergence of a 'closed European club' that neglects the European aspirations and the European nature of the Ukrainian state could strengthen pro-Russian sentiment among the population and further weaken Ukraine's fragile European hopes. Mrinska thus examines the consequences of EU enlargement on one of its largest new neighbours. Several question will be addressed, including: what influence will the enlargement process have on both foreign and domestic policy in Ukraine, and on different Ukrainian regions and communities? What should be done to soften the potentially negative effects of creating a new dividing wall on Ukraine's western border? Finally, the author will discuss how the EU's New Neighbourhood Initiatives targeted at Ukraine might help promote the harmonization of policies and actions on both sides of the common border.

Building on the arguments presented by Olga Mrinska, Tatiana Zhurzhenko discusses possible impacts of EU enlargement on regional cooperation beyond the EU's external borders and focuses on the 'new' borderlands between Russia and Ukraine. More specifically, Zhurzhenko will deal with new Ukrainian-Russian Euroregions that have emerged out of overlapping processes of European and post-Soviet integration. The central question here is whether the EU's New Neighbourhood Policy can, either now or in future, be understood as a geopolitical project in competition with Russian-led integration in the former Soviet Union. This chapter thus analyzes possible consequences of EU enlargement on the status of the Ukrainian-Russian border and perspectives of cross-border cooperation between the two countries. The problems of Ukrainian-Russian cross-border cooperation are more and more determined by the wider geopolitical context: this includes Ukraine's being torn between two competing integration projects – eastwards and westwards. Zhurzhenko asks how this situation might change with the institutionalization of 'Wider Europe' and a new status of Ukraine as a 'neighbour'. Presently the aim of the EU in regard to the eastern border of Ukraine is to monitor illegal migration and to help reduce migration flows by improving the efficiency of border control.

Will the enlarged EU be interested not only in security issues but also in supporting economic and humanitarian cooperation across this border?

Ilkka Liikanen and Petri Virtanen look at another important external border of the EU, that between Finland and Russia. This contribution examines the feasibility of regional institutional innovations in elaborating EU border policies and new instruments of cross-border cooperation. Within this context, Finnish-Russian border areas, and the Euregio Karelia in particular, are analyzed in terms of a case study of regional cooperation at pre-enlargement borders. In the first part of their study, Liikanen and Virtanen scrutinize different EU doctrines and instruments of cross-border cooperation (CBC) while addressing their roles in defining EU border policies. Attention will be drawn to the EU's Wider Europe strategy and possible consequences of the New Neighbourhood Initiative for cross-border interaction on the eastern borders of the EU. The historical and political development of the Finnish-Russian border region will also be scrutinized in order to clarify the existing preconditions for the adaptation of new CBC instruments. The main part of the study discusses Euroregions and their role in implementing EU border policies and focuses on the Finnish-Russian border and Euregio Karelia. In the concluding part of the essay potential roles of regional institutional innovations in adapting the EU's New Neighbourhood programme are outlined.

The third section of this book is primarily focused on everyday practices of cross-border cooperation and their regional embeddedness. Alla Skvortova begins this section by looking at how EU enlargement and EU policies are affecting the development of cross-border cooperation mechanisms, including Euroregions, between Moldova and Romania. In doing this, Skvortova provides a short overview of the history of relations between these two countries and traces, furthermore, the development of cooperation frameworks since Moldova's independence. Moldova and Romania have a long history of common ethnic, cultural and linguistic ties. As Skvortova points out, however, these commonalities have not always been helpful in overcoming political tensions. This is also true for the period after 1991, when Moldova regained its independence. However, Romania's invitation to join the EU accession process has helped Moldovan decision-makers realize that their country might benefit from Romanian support and expertise in their own negotiations with the EU. Within this context, Skvortova argues, the European Commission's Strategy on Wider Europe – Neighbourhood has had a very positive impact; this policy strategy emphasizes much more active cooperation between EU candidate countries and their neighbours in order to promote security and stability at the EU's future borders.

Béla Baranyi's essay summarizes results of an interdisciplinary research on the eastern borders of Hungary, the Hungarian–Ukrainian and Hungarian–Romanian borders in particular. As such, this chapter will deal with the present situation and the development prospects of border peripheries and speculates on the regional development contribution of different 'formations' of cross-border cooperation. Here, European integration, the connecting and separating character of the state borders and the new functions of the external borders of the European Union play an important role. Baranyi begins his analysis with an overview of the regional problems of these border areas, explaining how historical processes and postsocialist transformation have resulted in a 'periphery of the peripheries'. This is manifested by cross-border economic, social, cultural, institutional and ethnic relations. Another important regional issue is that of possible short- to mid-term effects of the Schengen border and new mechanisms of controlling flows across Hungary's borders with Romania and Ukraine. The study then deals with the potential development roles of different organizations of regional cooperation. These include already existing – and rather large – Euroregions, more recent 'small-scale' Euroregions, as well as 'micro-level' partnerships between communities. As an alternative to the existing and cumbersome interregional formations, small regional organizations on county or micro-regional level appear to be of increasing importance in promoting meaningful cross-border cooperation. In fact, argues Baranyi, there is evidence that the intensification of local cooperation is gradually contributing to the integration of economic spaces and urban market areas fragmented by borders and decades of 'non-cooperation'.

Gyula Szabó and Gábor Koncz elaborate on the previous contribution of Baranyi in looking at local-level aspects of cross-border interaction in the Hungarian-Romanian border region. The basis for this essay is fieldwork carried out in five settlements located near the Hungarian-Romanian border. These selected settlements are favourably situated from the point of view of cross-border interaction, they either have a border crossing at present, or a border crossing will be established in the near future. Due to Hungary's EU membership and changing geopolitical status, border-related issues and the question of economic development in regions along these borders have become of critical importance. The authors examine a particular aspect of this new relevance of borders in Hungary and neighbouring countries by documenting the experiences of the five border communities and persons living there. One of the essay's central objectives is to assess the efficiency and viability of local-level cooperation (that is, within Euroregional organizations). The authors also indicate the types of relations that have formed between these organizations and affected local governments and investigate to what degree the interests and strategies of these two levels of cooperation coincide in regard to future goals. Finally, reasons for both successes and failures of Hungarian and Romanian initiatives are enumerated.

The following contribution deals with Hungary's borders with Ukraine, where István Balcsók and László Dancs discuss patterns of legal and illegal employment of foreigners. Before 1989 cross-border labour-flows along the eastern borders of Hungary were but a minor issue. Borders were seen as highly restrictive zones and economic interaction was strictly regulated. With the disintegration of the socialist regime, borders became traversable and permeable, and regions along both sides of the borders once again established social and market interconnections. As a result, since 1990, cooperation between these border areas has become feasible. Intensifying cooperation has also had considerable impacts on local and wider regional labour markets. In this essay, prospects for effective cross-border cooperation in labour markets along Hungary's eastern borders will be scrutinized. In doing this the fact will be taken into consideration that these regions once functioned as organic economic and political units within Hungary during the period preceding the first world war, and that they today represent peripheral areas, having suffered negative effects of socialist isolation and the political truncation of their market areas. Adaptation to the new economic rules of the market and the EU have also proved a challenge for these regions. The authors therefore ask the question as to whether opportunities for efficient and meaningful labour market cooperation between Hungarian and Ukrainian communities in fact exist. Examination of this is all the more important, because Hungary's EU membership has made its labour markets more attractive to foreigners and thus could become a focal point of East-West labour flows. Already today cross-border employment, both legal and illegal, represents an important source of livelihood for families in Ukrainian Transcarpathia. However, despite the local importance of these labour market relationships, Balscók and Dáncs argue that it is unlikely that large numbers of foreign workers will come to this region; it is too peripheral and employment centres are at a considerable distance from each other. In addition the Ukrainian-Hungarian border section is now an external frontier of the EU and therefore subject to a strict application of Schengen criteria.

The final essay in this section deals with local and regional cross-border cooperation between Poland and Ukraine. The authors, Katarzyna Krok and Maciej Smętkowski, discuss emerging cooperation patterns within the new geopolitical situation in Europe. Polish membership in the European Union and transformation of the Polish-Ukrainian border into an external border of the EU, have unambiguously influenced relations and cooperation opportunities within the cross-border region. However, Krok and Smętkowski point out that fears of new exclusionary policies and a 'fortress Europe' syndrome threaten to encumber the local cross-border relationships that have developed between the two countries since 1991. In order to provide a picture of regional cooperation under these new conditions the authors analyze both the scale and the scope of these changes and the new role of different actors involved in cross-border cooperation. They also provide an overview of previous cooperation patterns and experiences and discuss future prospects of Polish-Ukrainian cross-border cooperation. In doing this, the socioeconomic and political conditions within which cooperation is emerging will be compared for both Polish and Ukrainian border areas. Based on this regional overview, local dimensions of cross-border cooperation will be analyzed and results of studies carried out in selected towns situated both on the Polish and Ukrainian side of the border will be presented and contrasted. This comparison allows the authors to provide deeper insights into practical aspects of different types of cross-border interaction and to identify the most important factors responsible for success or failure of public polices in this field.

The final section of this book is dedicated to the German-Polish border region, a former EU external border that is seen by some as a laboratory of post–1990 European integration. Grzegorz Gorzelak's essay is cautiously upbeat with regard to the legacy of cross-border cooperation in this region. While a 'great leap forward' in terms of economic development seems utopian, Gorzelak argues that cross-border cooperation in the German-Polish border region has had unquestionably positive results, results that perhaps do not enjoy as much publicity as they deserve. Here the author stresses that the benefits of cooperation must be understood in terms of learning processes in multilevel governance and in operating within contexts of EU public policy. Cross-border cooperation has empowered local governments in the German-Polish border region to act in a more forceful and self-assured manner and to grasp the potential advantages of EU integration. This has happened because they have been obliged to work with several levels of regional and national government, with different EU authorities and, ultimately, with each other. Interestingly, while Polish communities have been eligible for much less money from the EU than their German counterparts, the benefits of cooperation appear more tangible for the Polish side. Here, specific problems of East German transformation and a greater perception of East-West contradictions have tended to limit a sense of common European purpose among the citizenry.

In the final essay of this book, Hans-Joachim Bürkner is rather less sanguine about the prospects for effective regional co-development through German-Polish cooperation. However, here the focus is primarily on the Germany side of the border and the problems associated with social and economic transformation in eastern Germany. As Bürkner indicates, the development quandary of the German side of the border region is multilayered: high unemployment, a lack of local investment potential, depopulation and continuing marginalization threaten to make the region a permanent periphery. At the same time cross-border cooperation initiatives have flourished, spurred by the promise of EU funding for local and regional development projects. Nevertheless, viewed from more traditional perspectives on regional development, the results of Polish-German cross-border cooperation appear limited. Apart from a few visible success stories, such as joint university facilities in Frankfurt (Oder) and Slubice and the water treatment complex in Guben/Gubin, the border region remains very much divided. Entrepreneurial networks across the common border, for example, are weak and/or few and far between. However, at the same time, an active cross-border dialogue between public agencies, local government officials and NGOs appears to be emerging. What lessons can be learned from this rather ambiguous situation? In order to address this question the author focuses on contradictions between economic development priorities of the EU and state actors in Germany and Poland and actual regional needs. Fordist policies, with their emphasis on physical development and infrastructure, have characterized the main regional development doctrine on both sides of the border. What research increasingly suggests, however, is that the development of intraregional networks and a more visible campaign of regional identity-building is, for this specific region, of equal if not greater importance.

And Finally Some Acknowledgements

It is the hope of all those involved in the production of this volume that it will contribute in meaningful ways to the comparative study of borders and border regions.

We believe that border studies can contribute to understanding the relationships between territory, identity, citizenship, culture and governance. Furthermore, we also feel that multilevel analyses of how borders (understood here in the broadest sense of the term) influence human agency will help us comprehend how political community are being defined ('bordered') at different spatial scales (regional, national, supranational), and how these definitions are impacting on relationships between communities and states. It is, of course, up to the reader to determine the quality of our efforts. If this book should be seen to have at least partly contributed to the goals stated above then we should indeed be grateful.

At this point the editor would like to thank the following persons for their assistance, guidance and inspiration in putting this book together: Imre Nagy, Béla Baranyi and Zoltán Hajdú (who organized the 2003 Border Regions in Transition Conference in Hungary), David Newman, Olivier Kramsch, Pertti Joeniemmi, Heather Nicol, Rupert Hasterok, János Sallai, Silke Matzeit, Anssi Passi and Liam O'Dowd and Doris Wastl-Walter. Finally, thanks are due as well to the contributors of this book, their efforts and patience.

PART II
Borders and the Geopolitics of EU Enlargement

Chapter 2

Wider Europe: Geopolitics of Inclusion and Exclusion at the EU's New External Boundaries

James Wesley Scott

Interdependence – political and economic – with the Union's neighbourhood is already a reality. The emergence of the euro as a significant international currency has created new opportunities for intensified economic relations. Closer geographical proximity means the enlarged EU and the new neighbourhood will have an equal stake in furthering efforts to promote trans-national flows of trade and investment as well as even more important shared interests in working together to tackle transboundary threats – from terrorism to air-borne pollution. The neighbouring countries are the EU's essential partners: to increase our mutual production, economic growth and external trade, to create an enlarged area of political stability and functioning rule of law, and to foster the mutual exchange of human capital, ideas, knowledge and culture.

— The European Commission in 2003[1]

We cannot over-expand the EU. Europe has natural frontiers based on history.
— Helmut Kohl[2]

As enlargement of the European Union proceeds, relationships between the EU and neighbouring regions will rapidly change. May 2004 marked the establishment of the EU-25. In 2007, Romania and Bulgaria are scheduled to join. Turkey's membership within the next years is also a distinct possibility, the 'civilizational discourses' of Germany's Angela Merkel and others notwithstanding. As a result, the EU is extending the borders of its political community eastward to the former Soviet Union and southward towards Mediterranean and Middle Eastern regions. This will not only have long-term consequences for the new member states as they 'Europeanize' institutionally and apply EU regulations to their own borders, but it will also magnify the EU's macroregional geopolitical role.

As part of its goal to attain greater political stature and acceptance, and in recognition of its rapidly changing geography, the EU has redoubled efforts to define a sense of geopolitical purpose. This is taking shape in the guise of a geopolitical

---

1   European Commission (Commission of the European Communities 2003a, p. 3).
2   As quoted Brian Groom, 'Kohl voices EU expansion fears over Turkey entry', *Financial Times*, 21 November 2004.

doctrine that emphasizes the EU's stabilizing and democratizing role in the world system. Furthermore, the EU seeks to enhance its status within the world system through political and economic cooperation, rather than with military might. Additionally, the EU's 'soft power' approach is not only based on discourses of competitiveness and adaptability, but also on a notion of political integration that links economic, social, environmental and, increasingly, cultural issues. This has contributed to the emergence of a set of 'European values' that are now being projected onto both non-EU Europe and other regions of the EU's so-called near abroad.

One central element of this emerging policy, and one closely tied to the EU's nascent Common Foreign and Security Policy, is the promotion of friendly and effective working relationships with neighbouring regions. This involves new geopolitical conceptualizations of the immediate areas surrounding the present-day European Union. Emerging from specific local concerns over environmental safety, crime prevention, economic development and other issues, an implicit regional 'dimensionality' has made its imprint on EU politics at large. This dimensionality of EU geopolitics is reflected in the development of a European Neighbourhood Policy (ENP) as well as by discourses that support the notion of a 'Wider Europe'. The ENP, which will be operative by 2007, is primarily a strategy with which to rationalize and consolidate policies towards northern, eastern and southern neighbours, enhancing both the effectiveness and regional significance of the EU as a geopolitical actor. With the concept of Wider Europe, the EU aims to establish an ideational basis for political stability and economic growth within its immediate regional surroundings.

Above and beyond the exercise of 'soft power', the promise of a dimensional geopolitics lies in the establishment of informal and/or semi-formal regional dialogues between many different actors and across political, economic and cultural dividing lines. Because dimensionality informs the realpolitik of EU Neighbourhood affairs, this could lead to the establishment of comprehensive cooperation agendas involving more balanced partnerships. Such a symbiosis of formal and informal relations could, furthermore, furnish a workable alternative to strictly market-oriented (and increasingly neoliberal) readings of interstate cooperation and accommodate more appropriately heterogeneous economic and sociopolitical realities. However, as a EU geopolitical doctrine emerges, tensions due to simultaneous dynamics of inclusion and exclusion are very much in evidence. The idea of a Wider Europe is telling in itself: here, a sense of inclusion and belonging to a working political community is implied despite the fact that direct membership will be denied to many states that consider themselves very close to the EU. Furthermore, while the EU expresses a desire to avoid new political divisions, new visa regimes and other restrictions of cross-border interaction will possibly exacerbate development gaps between the EU-25 and non-EU states.

This paper will discuss challenges facing the Wider Europe and the emerging European Neighbourhood Policy and contradictions that characterize the EU's geopolitical ambitions. Particular attention will be paid to the post–cold war relationship to the 'North' (that is, with Russia) and the postcolonial cooperation context evolving with regard to the Maghreb, Middle East and other areas of the Mediterranean. In these regions partnership with the EU is subject to numerous obstacles that work against the sociopolitical basis for 'positive interdependence' the EU asserts to promote. The claim will be made, however, that dimensionality – for example, 'open' regional dialogues – can help bridge the political and economic discontinuities that exist between the EU and its neighbours by institutionalizing cross-cultural learning processes.

Wider Europe and Emergent EU Geopolitics

It is often stated that the European Union is much more than a mere free trade area and that it fulfils many state-like functions without being a state in the traditional sense of the term. In fact, the debate over the ultimate limits of EU jurisdiction over member states is incessant, often acrimonious, and will most likely continue well after a European constitution is agreed. One indication of the EU's increasing political weight are its ambitions to play a key geopolitical role within the world system. Despite the fact that consensus over the substance of its Common Foreign and Security Policy will be difficult to establish, the EU is nevertheless intent on defining a distinctive and coherent geopolitical doctrine that will help structure its relationships with other regions of the globe. In order for any meaningful EU geopolitics to emerge, the internal political community must be cohesive enough to allow agreement on the basic tenets of geopolitical doctrine, and a coherent set of guidelines and principles must govern interaction with those remaining outside that political community.[3]

From an official EU standpoint, the achievement of cohesion and coherence are the central goals of political integration and embodied in its 2001 white paper on European governance. Good internal governance and a responsive and democratic institutional architecture are, furthermore, understood to be prerequisites for promoting 'change at an international level' (Commission of the European Communities 2001b, p 26).[4] In more concrete terms this involves a process of community-building based on

---

3 Geopolitical regions can be defined as groups of countries that build associations around a set of common interests. The European Union, NAFTA, the Association of American States, MERCOSUR and APEC are all examples of cooperative institutions that integrate domestic and international policy agendas and 'collectivize' national interests (see Sum 2002). However, geopolitics be centred around a politically integrated community of states rather than a single nation or hegemon is a rather new feature of the global system and one that has seldom been tried in practice.

4 The White Paper continues along these lines with an appeal for greater geopolitical presence in order to strengthen the EU's sense of purpose: 'The objectives of peace, growth, employment and social justice pursued within the Union must also be promoted outside for them to be effectively attained at both European and global level. This responds to citizens' expectations for a powerful Union on a world stage. Successful international action reinforces European identity and the importance of shared values within the Union' (EU Commission 2001b, p. 26–7).

common rules and principles (including the so-called *acquis communautaire*) as well as adherence to a comprehensive set of political and ethical values (Antonsich 2002; Joeniemmi 2002).[5] Furthermore, the EU advocates a complex approach to regional development and cooperation in order to promote a sense of solidarity and socioeconomic cohesion. Here, economic, social, environmental and cultural issues are not only understood to be closely interrelated but also to be at the centre of peaceful interstate relationships and a prosperous EU (see Barnier 2001; Prodi 2001).

The 'deepening' and 'widening' of the EU has thus elicited a variety of spatial strategies that cut across traditional nationally-oriented development practice.[6] In terms of continental development, the European Spatial Development Perspective (ESDP) envisions a strategic European space that is networked, flexible, competitive, but at the same time cooperative in the solution of common problems (Commission of the European Communities 1999). Similarly, local and regional cross-border cooperation and other forms of societal interaction between states are seen as important aspects of EU integration and have acquired considerable political significance as a mechanism for deepening relations with non-EU neighbours (O'Dowd 2003; Scott 1999).

This all signals, at least theoretically, a departure from traditional forms of international relations and represents a source of European 'uniqueness' within the world system. In fact, Björn Hettne (1999), Pertti Joeniemmi (1996) and other scholars see the EU in the vanguard of a post–cold war geopolitical 'order' – an order characterized by multipolarity and transition, a move towards comprehensive cooperation agendas, multi-actor policy arenas and the development of a transnational civil society.[7] In the words of Hettne (1999), a 'new' regionalism is at work here that

---

5   A notable element of the Treaty on European Union (Maastricht Treaty) was the introduction (in Articles 8–8e) of legal and conceptual elements of formal European citizenship into an integration process hitherto characterized primarily by economic issues. Going a step further, one of the implicit goals of the 1998 Treaty of Amsterdam is the promotion of a European public sphere through the establishment of common (that is, unifying) constitutional principles and intergovernmental processes. These arrangements are also intended to support the definition and acceptance emergence of common values such as in the area of human rights, women's rights, democracy, etc. (Pérez Diaz 1994).

6   The terms 'deepening' and 'widening' are characteristic of post-Maastricht EU discourse and are used to convey a sense of closer political and economic integration as enlargement proceeds.

7   In a new regionalist reading, a 'post-Westphalian' security regime would not be centred on the interests of individual states and on balances of power. It would rather be based on a recognition of interdependence between nations and the necessity of a much wider political agenda of development in order to stabilize the global geopolitical system. Notions of hegemony backed up by military and economic might are thus foreign to this 'security complex'. A post-Westphalian political and security perspective would focus instead on non-exploitational interdependencies (partnerships), environmental issues and, hence, on a 'collectivization' of national security (see McGrew 2000).

can be defined as a 'multidimensional process of regional integration which includes economic, political, social and cultural aspects' (p. 17). In stark contrast to the EU's approach, the North American Free Trade Area (NAFTA), Mercado Común del Sur (MERCOSUR) and Asia Pacific Economic Cooperation (APEC) remain focused on the geoeconomics of regional cooperation and open markets.

Consequently, the EU sees itself as a major stabilizing factor in the world system, not only because of its economic weight but because of its 'ideational projection'. Antonio Guterres (2001), the former Portuguese prime minister (now president of the Socialist Internationale) has stated in prosaic terms that 'the European Union – as the only true organized regional space in the world – plays a fundamental role in the building up of the new political architecture and needs to be strengthened further. Indeed Europe, through its integration process, has been able to act not only as an element of balance in the international relations, but also as an inducing factor leading to the strengthening of other regional blocs' (p. 8).' Guterres has emphasized, furthermore, the 'fact [that] only with strong regional organizations, preserving each of them their own social and political models, will we be able to build up a multipolar world to avoid a savage and uncontrolled globalization which, most probably, would open the way to a globalization of poverty and to a decrease, at its lowest level, of economic and social rights all over the world' (ibid.).

Guterres's statement highlights the main ideological positions with which the EU is defining its geopolitical role. However, it is only recently that this 'doctrine' has acquired more concrete expression in policy terms. In its widely publicized March 2003 communication 'Wider Europe – Neighbourhood: A New Framework for Relations with our Eastern and Southern Neighbours', the European Commission has defined contours of an initiative that aims to 'enhance relations with its neighbours on the basis of shared values', to 'avoid drawing new dividing lines in Europe' and to promote 'stability and prosperity within and beyond the new borders of the Union' (Commission of the European Communities 2003a, p. 4). Furthermore, this initiative 'is to be based on a long term approach promoting reform, sustainable development and trade' (ibid.). Importantly, the EU Commission emphasizes interdependence as a central geopolitical principle: 'enhanced interdependence – both political and economic – can itself be a means to promote stability, security and sustainable development both within and without the EU. The communication proposes that the EU should aim to develop a zone of prosperity and a friendly neighbourhood – a "ring of friends" – with whom the EU enjoys close, peaceful and cooperative relations' (ibid.).[8] These concepts have been elaborated in a Strategy Paper and a

---

8 The 'ring of friends' metaphor was introduced by Commission President Romano Prodi (2001): 'With globalization and the creation of a trans-national civil society, the Union's external relations can no longer be distinguished from its internal development, particularly when it comes to our neighbourhood. Instead of trying to establish new dividing lines, deeper integration between the EU and the ring of friends will accelerate our mutual political, economic and cultural dynamism.'

concrete Proposal for a European Neighbourhood Policy (ENP), both presented in 2004.[9]

A continuing (or 'rolling') process of enlargement is seen as unsustainable (Wallace 2003), but exclusion from the EU will be compensated for by closer cooperation, the promise of free trade and more effective regional assistance programmes. Günther Verheugen (2003), EU Commissioner for Enlargement, has defined this initiative in the following way: 'The creation of a 'wider' Europe means the creation of a common economic and social space where all countries can potentially have access to the internal market. It also means opening up and cooperating more intensively in a very broad range of EU internal policies – from transport to the environment, from justice and home affairs to security and defence'. (p. 5) Attempts are thus under way to harmonize financial requirements and reduce administrative barriers to bilateral and multilateral activities with non-EU countries. As a consequence, the ENP will replace the various cooperation mechanisms operating since 1990. By 2006, according to the EU's provisional timetable at least, these programmes will be subsumed into country-specific and regional action plans that will allow a genuine cross-border financing of projects.[10] This would signify a marked change from the strict adherence to territoriality principles that has characterized EU policies.

Wider Europe and it policy-driven counterpart, the ENP process, underscore the extent to which enlargement is transforming relationships and the significance of national borders between the expanding EU and its neighbouring states and regions. As a result, the EU is increasingly challenged to positively influence developments in its immediate 'neighbourhood', promoting stability and economic development as well as accommodating different geopolitical sensibilities and scales of cross-border interaction with regard to neighbouring regions. If we scrutinize European policies and discourse of the above variety more closely, it becomes apparent that the EU is, in effect, attempting to 'Europeanize' not only the political community of states that it represents but also the greater regional space around it. Within the last ten years EU policy initiatives, for example, have assumed a much more active role in determining institutional conditions for local and regional cross-border cooperation.

Several questions thus emerge with regard to the possible challenges that 'Europeanization' will face. Above all, it is unclear to what extent the geopolitical concept of Wider Europe and the ENP instrument can achieve the ambitious goals set out in the September 2004 proposal. Can they serve to develop partnerships and improve conditions for multilevel, cross-border interaction between the EU and it surrounding regions? In order to properly address this question it is necessary to focus on the simultaneous dynamics of inclusion and exclusion that have emerged with regard to the 'near abroad' of Europe, regions closely tied for centuries to the

---

9   See Commission of the European Communities 2004a and 2004b.

10   Most prominent of the EU initiatives are INTERREG (regional development aid for EU border regions), PHARE (development assistance for Eastern European countries), TACIS (technical assistance for the Community of Independent States) and MEDA (cooperation with Mediterranean states).

countries of Western and Central Europe. Representations of European space, such as those proposed in the ESDP, are aimed at promoting a sense of European cohesion. At the same time they often convey a sense of core-periphery relationships, not only within the enlarging EU itself but also between the EU and regions adjacent, such as the former Soviet Union and Mediterranean states (see Scott 2002).

It is in this context that the notion of 'dimensionality' is often applied to the EU's approach to regional partnership. EU support of transnational and cross-border cooperation is increasingly informed by specific regional perspectives that have emerged as a result of the EU enlargement process (see, for example, Cimosewicz 2003; Decker 2002; Mazur 2002, Reut 2002). At present four partially overlapping mesoregional contexts (or dimensions) can be discerned within Wider Europe and NNI: 1) a 'northern' dimension directed towards the Baltic and Barents Sea areas, 2) an 'eastern' dimension that involves Russia, Belarus and the Ukraine, 3) a Mediterranean geopolitical 'partnership' encompassing 12 non-EU states and 4) a specific Balkan dimension concerned with peaceful development in Albania and parts of the former Yugoslavia.

The Northern Dimension: Development of a Post–Cold War Geopolitical Perspective

The Northern Dimension (hereafter ND) is a mesoregional geopolitical strategy that draws much of its impetus from post–cold war rapprochement in the Baltic Sea Region (BSR).[11] Its beginnings are to be found in regional attempts to manage post–cold war economic, political and social transformations as well as deal with the environmental problems of the Baltic and Barents Seas. As an early response to the collapse of the Soviet Union and the changing geopolitical situation in the European 'North', the Norwegian and Finnish governments, as well as other states in the region, began to take advantage of new opportunities for political dialogue. Central to this process was and remains the facilitation of new development perspectives for the Baltic and Barents Sea regions through multilateral cooperation and, above all, through the development of trust-building partnerships with Russia. This became all the more vital in conjunction with EU enlargement (Finland and Sweden became EU members in 1995) and NATO's expansion to include Poland and the Baltic States.[12] The 'official' proclamation of the existence of a 'Northern' Dimension of the EU came two years after Finland and Sweden joined the EU in 1995. In a now famous speech delivered at Rovaniemi, during a conference on Barents Sea Region cooperation, the Finnish prime minister, Paavo Lipponen (1997), appealed for an

---

11 The Baltic Sea Region comprises 11 states in Scandinavia and central and eastern Europe: Norway, Denmark, Germany, Sweden, Poland, Belarus, Kaliningrad (Russia), Lithuania, Latvia, Estonia, Finland and the Karelian and Russia (St. Petersburg and Karelian Districts).

12 The Baltic States – Lithuania, Latvia and Estonia – regained their independence after the 1991 collapse of the Soviet Union.

EU strategy addressing the particular regional problems of its northern member states and Northwestern Russia. The principal message here was to sensitize the EU as a whole to a new regional security perspective within the wider European context, one emphasizing serious environmental issues, nuclear safety, crime prevention and minority rights rather than questions of national defence. In contrast to NATO's controversial expansion eastward, traditional security issues have thus been subsumed into the ND's comprehensive regional agenda – an agenda that emphasizes functional international cooperation and the strengthening of institutions of a democratic civil society (Joeniemmi 1999).

Indeed, the EU's role in the Baltic Sea Region and the wider 'North' has increased rapidly. The conclusion in 1992 of a Partnership and Cooperation Agreement between the EU and Russia provided vital support to international cooperation efforts in the region. The EU has, furthermore, emphasized support of a meaningful political partnership and a 'positive, broad and ambitious economic agenda' with the Russian Federation.[13] Above all, the notion of 'positive' (that is, non-exploitative) interdependence has been held up as the basis of future partnerships between the Baltic Sea Area, the EU and Russia (Lipponen 2002). In December 1997 the Luxembourg European Council formally agreed to introduce a northern dimensionality into its internal and external polices.

A further step towards strengthening ND was taken with the commissioning by the 1999 Helsinki European Council of an 'Action Plan' with regard to external and cross-border policies of the EU. This Action Plan was adopted a year later at the Council's meeting in Feira, Portugal, and remained in effect until 2003. A Second Action Plan has been established for the period 2004–2006. With these action plans a substantive agenda has been defined that identifies sectors where cooperation is most necessary. These sectors include environmental protection, nuclear safety and nuclear waste management, business development and investment, cooperation in the energy sector (the region has considerable gas and oil resources while the EU's energy needs are likely to increase after enlargement), transportation issues, the improvement of border crossing facilities, crime prevention (in areas characterized by wide gaps in living standards!), public health, social programmes, telecommunications, human resources development, protection of indigenous peoples of the North, and finally, the solution of geopolitical and economic development problems associated with Kaliningrad's enclave status (Council of the European Union 2000 and 2001). Additionally, a Northern Dimension Environmental Partnership (NDEP) was started in 2001 in order to enhance the ecological element of regional cooperation. This

---

13 Quoted from a letter written by Pascal Lamy, EU Trade Commissioner, and Chris Patten, EU External Affairs Commissioner, to the *Financial Times* (17 December 2001). Furthermore, the EU pursues a Common Strategy, agreed by the European Council in June 1999, in order to develop a comprehensive political partnership with the Russian Federation as shown, among other things, by EU support of Russian accession to the World Trade Organization, which signals a desire to positively influence institutional reform in that country.

partnership, coordinated by the European Bank for Reconstruction and Development (EBRD), targets in particular so-called hot spots in the Kola Peninsula and other parts of Northwestern Russia and seeks to work with International Finance Institutes in order to fund regional cooperation projects (Haukkala 2001a). In total the NDEP's Steering Group has, in a 'bottom-up' selection process, listed environmental protection and nuclear safety projects of an estimated aggregate cost of almost € 2 billion (Steering Group of NDEP 2002).

Much attention is paid in ND discourse to the municipal and regional levels as these are seen as the basic units where civil society interacts transnationally (Browning 2001) and as a necessary requisite for achieving 'positive interdependence' (Sergounin 2001). Euroregions, such as the Finnish-Russia Euregio Karelia, have been established in the Northern Dimension area since 1995 in hopes of capturing economic and social benefits from cooperative projects (Eskelinen 2000, Reut 2000). The Euroregion Karelia, for example, announces itself as a regional player within the Northern Dimension.[14] In addition, a number of project-oriented Baltic Sea networks have established a precedent for international civil society activism. Presently dozens of initiatives involving cities, regions, chambers of commerce, universities, national governments, NGOs and other actors are either under way or in preparation (Scott 2004). Their main objective is to introduce alternative regional perspectives into the strategic orientations of the EU and nation-states.

As a result of this track record, ND will form a basic element of the EU's New Neighbourhood Initiative (NNI) – and has, in fact, already been taken up as an element of the EU's Common Foreign and Defence Policy. In this regional context, a precedent for multilevel and multifaceted cooperation has already been established, and it is only logical that the EU build upon this heritage. One stumbling block to cooperation, namely the question of Russia's Kaliningrad enclave and the freedom of its citizens to travel freely between Kaliningrad and greater Russia once the territory has become enclosed within the EU as of May 2004, appears to have been successfully removed. Thanks to the diplomatic initiative of the EU a compromise solution was agreed in November 2002, providing for special travel documents and transit corridors through Lithuania.

However, despite the considerable momentum behind the ND and its perceived ability to advance regional environmental agendas, true regional partnership – particularly between Russia and the EU – remains elusive. This problem will most likely remain a major obstacle to the realization of a 'Wider Europe' (at least as envisaged by the EU) for quite some time. Reasons for the ND's difficulties in bringing Russia closer to the EU have been varied. Funding, for example, has always been a weak aspect of ND, and it has been deprived of resources – from the very beginning in fact – so as not to elicit opposition from the 'South'.[15] Indeed, initial

---

14 See the website <www.karjala-interreg.com/euregio/eng/>.

15 Up to the present ND has operated without a specifically dedicated source of EU funding. Instead, project facilities have had to be constructed through a variety of means, often quite independently, in each of the strategic areas defined in the Action Plan. Generally,

hostility towards the ND, particularly from the French and Spanish governments, was grounded in fears that a Northern agenda could detract from the Euro-Mediterranean partnership and the Barcelona process (Terva 1999).[16] On the other hand, the NNI will theoretically provide a major source of funding once regional programmes are agreed, and this could well cease to be a major issue. Much more problematic is the sociopolitical basis of this partnership. Contradictory regionalization logics, cultural difference and simultaneous processes of 'exclusion' and 'inclusion' represent a persistent barrier to more open cooperation.

The Russian situation is one of a formerly centralized federation that is fragmenting due to a variety of processes weakening Moscow's effective control of regions, thus reducing the prestige of federal institutions (Herd 2001). As a result, sovereignty issues have remained highly sensitive, with the maintenance of national integrity the main policy concern of the Federation. Russia's uncompromising stance on Chechnya has, furthermore, alienated many in the EU who had wished to see the development of a more civil and democratic Russia. But sovereignty is not merely a Russian preoccupation and beyond it there are more subtle issues at play. Iver Neumann (1999) has pointed out that regionalization processes (in his specific case, the definition of a post–cold war Baltic Sea space of cooperation) involve selective scalings of 'we-and-them' categories based on different levels of regional purpose. By drawing parallels with other contemporary and competing forms of regionalization, the EU in fact conceptualizes itself as a higher form of political cooperation between states that is not only institutionally sophisticated but better attuned to global issues and human and social rights.[17] However, as fas as the building a new strategic development and security partnership between Russia and the EU is concerned, this could be interpreted in terms of 'modernist' core-periphery geopolitics. As Catellani (2002) states this relationship 'more closely [resembles] aid-like dynamics [rather than] a partnership based on a balanced exchange of resources' (p. 17). Russia, despite its internal weaknesses and fractures, and its loss of international political stature, is 'psychologically unprepared for the role of periphery (Benediktov 2002:1)' and can be expected to react negatively to EU condescension.

---

this has meant the use of already existing EU funding mechanisms such as INTERREG and PHARE and securing support from International Financial Institutions (IFIs) such as the EBRD, the Nordic Investment Bank (NIB), and the World Bank Group. Smaller public finance institutions, such as the Nordic Environment Finance Corporation (NEFCO) and the Nordic Project Fund (NOPEF), are involved as well.

16 Jukka Terva, almost certainly a pseudonym meaning 'jack-tar' in English, has engaged in intentionally polemic but scholarly internet debate on EU policies as they affect the Baltic Sea Region.

17 A European interpretation of North American rationalization, for example that of Zaki Laidi (1998), might stress that NAFTA is less a macroregional political community than a project of continental economic regulation focused on and dominated by US-American economic power and characterized by highly asymmetric relations between its member states; hence a lack of political institutions with which to widen transnational cooperation, a lack of a truly regional focus and, perhaps most fundamentally, a lack of a focus on society.

As is the case with the overlying geopolitical situation of the European North, the pervasive centrality of the EU-Russia relationship impacts on the local and regional levels as well. Here, the asynchronous and diverging regionalization logics discussed above remain persistent obstacles to more effective 'EU–non-EU' cooperation. Perceptions of this asynchronous regionalization revealing. As one prominent Finnish observer, Hiski Haukkala (2001a) notes, Russian regionalization processes and the cooperation developing between Russia and the EU are strongly influenced by a 'nation-building' mentality and emphasis on the integrity of state borders reflecting a realist and/or 'modern' geopolitics'. This is seen to be incompatible with the European notion of shared sovereignty and positive interdependence.[18] Furthermore, the present crisis in Chechnya and EU opposition to Russia's handling of this regional conflict serve only to exacerbate the perceptions of sociopolitical, cultural and systemic difference.

By the same token, however, resentment against a 'normative divide' due to a perceived discrimination of Russia and EU favouritism towards other eastern European countries is widely felt among Russian elites (Light, White and Löwenhardt 2000). The EU appears to expect a 'westernization' of the East in the sense that Russia develop a democratic society sharing its core values. However, Russia perceives the EU as a neighbour in economic terms rather than a strategic political partner and feels rather less compelled to adopt values central to the EU (ibid.). As a result, asymmetries of interests and cooperation perspectives are a severe problem, for example, for Finnish-Russian Euroregions and the Estonian-Russian border situation. Local projects meant to unite communities along this 'divide' suffer from the paternalistic (if not downright patronizing) decision-making style of Brussels that appears to differentiate between 'western' and 'eastern' mentalities (Haukkala 2001b). The message conveyed is clear: Europe outside the EU, particularly the 'East', cannot be trusted (Cronberg 2001).

The EuroMed Partnership: A Postcolonial Development Dialogue

The Mediterranean region presents a geopolitical challenge no less complex than the EU's partnership with the former Soviet Union. Because of the region's extreme heterogeneity and fragmentation, EU relations with these states have been largely bilateral and lacking a coherent regional perspective. In addition, the intensity of relations maintained by Mediterranean states with the EU has also been highly varied.[19] The Maghreb states of Morocco, Algeria and Tunisia were former French colonies and remain highly dependent on the European economy. The other states (that is, Egypt, Jordan, Lebanon, Syria, and Israel as well as the Palestinian Authority) make up a complex geopolitical constellation where economic ties, while important, tend to be overshadowed by security issues of larger international impact. Here, the EU is trying to take a leading role in promoting peace and stability and thus vying

---

18 Haukkala is affiliated with the Finnish Institute of Foreign Affairs.
19 See Commission of the European Communities 2003a.

with the United States for geopolitical relevance. Libya, once isolated politically, is now being integrated into the EU's Mediterranean dialogue. Finally, there is the often ambiguous status of Turkey, a very close partner of the EU and – the heated debate on the matter notwithstanding – a possible future candidate for EU-membership.

France, Spain and Italy have been principal protagonists in developing a 'southern' (that is, Mediterranean) geopolitical dimension for a number of reasons. The urgency of developing a cooperation agenda with the poorer states of the Mediterranean results partly from the very close economic dependency of the Maghreb and other countries in the region on the EU, geographic proximity and strong postcolonial ties. Indeed, pressures originating from illegal immigration and illicit trade have highlighted the sensitivity of the EU's southern borders. Furthermore, and as mentioned above, fears that enlargement eastward would divert attention and resources from the specific problems of the Mediterranean area created pressure for a more decisive community approach to the region. In response to these pressures the EU has, since 1995, embarked on a policy of Euro-Mediterranean Partnership with the 12 states of the Mediterranean (Aghrout 2000).[20]

The Euro-Mediterranean Partnership (EMP), was agreed by the foreign ministers of the EU and the 12 Mediterranean partners at Barcelona in November 1995 (for this reason the EMP is often referred to in general terms as the 'Barcelona process'). The guiding principle behind the EMP is a recognition of the fact that the promotion of democracy and stable societies in the southern Mediterranean will require a more active role on the part of the EU and a dialogue where not only economic development (such as the creation of a free trade area) but also broader political and cultural dialogue are prioritized.[21] The three primary goals of EU policy towards the Mediterranean are enshrined in the declaration adopted by the 27 countries that took part in the Barcelona Conference of November 1995;[22] these are:

1. to create a common Euro-Mediterranean area of peace and stability based on a number of fundamental principles, including the respect of human rights and democracy;

---

20 The states involved in the 'Southern' dimension, or Euro-Med Partnership are: the 15 EU members and the 12 Mediterranean non-EU member countries (so-called MNCs) of Algeria, Cyprus, Egypt, Israel, Jordan, Lebanon, Malta, Morocco, the Palestinian Authority, Syria, Tunisia and Turkey. Mauritania, the League of Arab States and the Arab Maghreb Union (AMU) were also invited to attend the 1995 Barcelona conference. Interestingly, Wider Europe and NNI do not address Turkey (along with Bulgaria and Romania) as the prospect of future EU-membership appears to exclude it from the 'near abroad' category.

21 To quote Chris Patten (2001): 'The objectives agreed on at Barcelona remain fundamentally valid and even increasingly relevant: working together for *peace and stability*; creating *shared prosperity* through establishing *free trade* and providing the *economic and financial assistance* to meet the challenges which that implies; and helping to improve *mutual understanding and tolerance* among peoples of differing cultures and traditions.' (p. 1; emphasis in the original)

22 Original text available at <europa.eu.int/comm/external_relations/euromed/bd.htm>.

2. to construct an area of shared prosperity through an economic and financial association which will favour the progressive introduction of a free trade area;

3. to commence wide-ranging action designed to build closer contacts between the different peoples of the region through a social, cultural and human partnership aimed at encouraging understanding between cultures and exchanges between civil societies.

The partnership is based on association agreements negotiated both bilaterally between the EU and its 12 partners, and multilaterally through regional bodies, such as the Euro-Mediterranean Committee and the regular Senior Officials' Meetings on the Political and Security Dialogue. Significant results of the Barcelona process have been the creation of a funding mechanism (the MEDA initiative) for cooperation projects between the EU and Mediterranean states. During the 1995–99 programming period MEDA was allotted € 3,435 million (making up the lion's share of the approximately € 4.4 billion dedicated to cooperation between the EU and its Mediterranean Partners). The present 2000–2006 phase has seen an increase in funding, with MEDA receiving € 5,350 million. Apart from direct EU grants, the European Investment Bank (EIB) has made substantial loans to Mediterranean states; total loans provided during the MEDA I and II phases will likely total over € 11 billion. Other results of the Barcelona process have been a Common Strategy for the Mediterranean Region adopted in June 2000 and, in similar fashion to the Northern Dimension, an Action Plan agreed in April 2002.

As is the case with the ND, economic development and positive interdependence are seen as a means of promoting democratization, strengthening civil society, stemming the tide of illegal immigration, combating crime, etc. Unsurprisingly, free trade is the concrete political goal with which 'mutual prosperity', multilateralism and regional integration are to be achieved. The EMP prioritizes the gradual establishment of a EuroMediterranean Free Trade Area between the EU and its partners, on the one hand, and among the partners themselves, on the other, by the target date of 2010; this process is accompanied by substantial financial assistance from the EU (principally the MEDA programme) and through European Investment Bank loans to promote economic transition and to help the partners meet the social and economic challenges implied by these changes (Commission of the European Communities-External Relations 2003, p. 3). This free trade zone, if implemented, will link the EU-25 with the 12 Mediterranean partners creating one of the world's largest geoeconomic regions.[23]

---

23 The Euro-Mediterranean Free-Trade Area foresees free trade in manufactured goods and the progressive liberalization of trade in agricultural products and, eventually, services. It is being gradually implemented through Euro-Mediterranean association agreements. Association agreements have already been established with Turkey, Cyprus and Malta and include the creation of customs unions. The Association agreements also deal with cooperation in political, economic, social and cultural matters, justice and home affairs. The agreements

The EMP also includes social and cultural components. In this context the promotion of cooperation between representatives of civil society is seen as critical to greater understanding and intercultural dialogue. Since the establishment of the EuroMed Partnership (EMP) at the 1995 Barcelona Conference, fora representing civil society organizations have become a prominent feature of these regional summits.[24] The EuroMed Civil Forum, as it is known, deals with a variety of issues related to democracy, economic development, human rights, etc. So-called decentralized cooperation programmes have been introduced within the scope of MEDA since 1995 to stimulate participation of civil society organizations (Med-Urbs, Med-Campus, Med-Media) in small-scale projects.

The goals set out by the Barcelona process are virtually identical to those expounded in the Wider Europe communiqué of the European Commission. As such, the EMP represents a central dimension of Wider Europe and the New Neighbourhood Initiative. Conspicuously, the Mediterranean dimension of the EU's emerging geopolitical strategy is heavily influenced by economic agendas and the management of the negative 'externalities' of market expansion. In contrast to the Northern Dimension, with its strong ecological and sociopolitical focus, the achievement of a free trade area is central to the EU's drive for regional cooperation in the Mediterranean. Increased regional trade and investment, a general goal of Wider Europe, is seen as a mechanism for combating poverty and thus reducing pressure on the EU to accommodate immigrants from the region. At the same time border-related issues loom large and the EU demands 'shared responsibility' for combating illegal cross-border activities and 'common security threats'.

As a result, the problem of reconciling inclusionary partnership with exclusionary security policies is perhaps even more pronounced in the Mediterranean context than with respect to Russia and eastern Europe. The Civil Forum, while welcoming the development of multilateral dialogue between the EU and the greater Mediterranean region, has, for example, been quite critical of the EMP. One source of dissatisfaction is the perceived impotence of the EU and the EMP in positively influencing the search for peace in the Middle East (EuroMed Civil Forum 2003). Furthermore, economic and security agendas are seen to be focused primarily on a rather one-sided notion of free trade and on controlling illegal immigration, marginalizing the very sociopolitical elements that EU discourse so vocally promotes (for example, democratic progress, sustainable development and human rights).[25] Culturally and

---

can be suspended if essential elements (such as human rights and democracy issues) are not fulfilled.

24 EuroMed summits have been held in Barcelona (1995 and 2005), Malta (1997), Naples (1997 and 2003), Stuttgart (1999), Marseilles (2000), Brussels (2001) and Valencia (2002).

25 The Spanish newspaper El País, on 26 October 2003, has devoted several pages to immigration issues; the main article entitled 'El reto de la inmigración divide a Europa' (The challenge of immigration divides Europe) illustrates the inability of the EU to develop a coherent policy with which to normalize the inflow of immigrants. Also lacking are policies of cooperation with countries such as Morocco that either stimulate immigrants to return or

small group–oriented programmes (the so-called decentralized programmes) have, unfortunately, been poorly managed and the EU will most likely discontinue these programmes.[26]

Political problems and the shaky Middle East peace process have, of course, had a negative impact on the ambitious agenda defined by the EuroMed Partnership. Similarly, to many observers from the region, the EU lacks credibility because it appears to be more interested in pursuing its own economic gain than seriously promoting human rights.[27] These concerns have also been voiced by the EU's Economic and Social Committee (ESC) who in an evaluation of the EMP stated that 'the weakness of the Euro-Mediterranean partnership lies precisely in the emphasis on its political and economic content, with social content seriously lacking' (Commission of the European Communities-ESC 2002, p. 119). Perhaps more damagingly, the ESC has criticized the eurocentrism of the Free Trade Zone project. In what could be interpreted as thinly veiled neocolonialism, the FTZ envisaged by the EU would privilege industrial sectors in which the EU is competitive while downplaying the role of agricultural trade, traditionally of great importance to Mediterranean countries.[28] Viewed perhaps somewhat uncharitably, the MEDA programme is basically a form of development aid that is compensating for the disappearance of trading preferences while promoting economic transition.[29]

---

that prepare and train prospective migrants for employment in the EU. As a result, several EU states have taken matters in their own hands. Italy, for example, has drawn up agreements with Tunis and Albania that have succeeded in reducing (illegal) immigration from these countries.

26 Chris Patten (2001, p. 5), EU Commissioner in charge of External Relations, has stated that such 'microprojects' are not cost-effective and tie up too much time and energy.

27 Comments made by Walid Kazziha, American University in Cairo, at the April 1998 conference 'Is the Barcelona Process Working?' (Philip Morris Institute 1998, p. 30).

28 According to the ESC in its recommendations for the improvement of the EuroMed Partnership: '[I]t is necessary to defend the idea of converting the FTZ plan into a real "common market" which would include all the goods covered by the sectors where the South is competitive, i.e. essentially the agricultural sector. Such a market presupposes a very significant supporting policy (technical and health-related specification of products, modernization and re-orientation of certain sectors, modernization of firms involved in the processing and marketing of agricultural products, etc.) as a second key objective.' (Commission of the European Communities-ESC 2002, p. 121)

29 One indicator of this is the bilateral, rather than regional, nature of the programme. During the period 1995–99, an overwhelming 86 per cent of the resources allocated to MEDA were channelled bilaterally to Algeria, Egypt, Jordan, Lebanon, Morocco, Syria, Tunisia, Turkey and the Palestinian Authority. Only 12 per cent of total resources were devoted to multilateral and/or regional cooperation activities. For the period 1995–99, grants supported four major areas of activities: 1) support to structural adjustment (15 per cent of total commitments) 2) support to economic transition and private sector development (30 per cent), 3) classical development projects, such as education, health, rural development (41 per cent), 4) regional projects (14 per cent).

However, the total level of compensation falls well short of medium-term losses of revenue these countries, and particularly the Maghreb states, face (Aghrout 2000).

In summing up, the Barcelona process, as an element of Wider Europe, appears to lack a strategic vision for the development of regional partnership. The EuroMed Civil Forum (2003) has complained that:

> The Partnership's stress upon security/stability imperatives has [...] raised the question of whether the EuroMed process is geared to contain conflicts rather than to facilitate sustainable, just solutions to these conflicts. Furthermore, the definition of the EuroMed region appears artificial, driven by the geo-political interests of European states, which serve only to increase the divisions in the region instead of creating a more equitable playing field. (p. 4)

In a similar vein, Georgios Papandreou, Greece's alternate minister of Foreign Affairs, has stated that 'the northerner's neocolonial attitude has yet to be replaced by an equitable, substantive relationship that is much needed to tackle the problems that spill from one shore to another' (Philip Morris Institute 1998, p. 7).

Thus here, as is the case in Europe's 'North', the interests and perceptions of the EU dominate. Free trade favours European exports. Issues of illegal immigration and security remain central to the partnership and negatively influence social and cultural elements. With much of the EU's attention and energies focused on the challenges of eastward enlargement, the Civil Forum even senses a relative loss of interest in the EMP, except to the extent where the recent 'war against terrorism' is concerned. Consequently, there appears to be little interaction between the principal actors involved in the EMP and organizations of civil society. Whether or not this situation will continue can only be answered in the mid-term. Chris Patten (2001), EU Commissioner for Foreign Affairs, has at any rate signalled the necessity for a more sensitive approach to partnership with southern Mediterranean states:

> The Mediterranean is our 'near abroad' on our Southern flank. Thus, it should enjoy a special place in our external relations. The EU and its Mediterranean partners share many common interests: from trade and investment, through safeguarding the environment and energy supply, to maintaining regional peace and stability. The partnership established at Barcelona in November 1995 [...] recognized that common objectives and common interests need to be addressed in a spirit of co-responsibility leaving behind the more 'patronizing' approach which often marked our policy in the past. (p. 1)

Conclusions

This discussion of dimensionality and the Wider Europe initiative indicates that EU geopolitics in the sense of a coherent foreign policy and security doctrine are gradually taking shape. This is, however, as yet a project very much 'under construction'. The two regional perspectives focused on (that is, the Baltic Sea and the Mediterranean) give evidence of the seriousness of Brussels' attempts to develop a new quality of societal and political interaction with the states of these two regions. With time it is likely that the EU will develop much more flexible and attractive structural policies with which to facilitate cooperation with the 'East' and 'South'.

Dimensional aspects of EU geopolitics, however, also serve to highlight the basic weaknesses and contradictions of the Wider Europe strategy. A basic premise of the European Commissions' notions of a 'wider' Europe is that 'positive' political and economic interdependence will form the basis of regional partnerships with the new and enlarged EU political community. However, this idea does not sufficiently address the central problem of asymmetry, both in terms of economic and political development, and of imbalances of power. Partnership is rarely possible where rich and powerful states are able to dictate the basic conditions of cooperation to others. While power asymmetries are a crucial issue, they are exacerbated by inconsistencies in the EU's own geopolitical rhetoric. European discourses of positive interdependence underlie the idea of 'Neighbourhood' while free trade and open borders are upheld as necessary for economic partnership. At the same time stricter regulation of the EU's external boundaries threatens to limit the extent to which transnational civil society and sociocultural cooperation can develop. In addition, the EU is seen as pursuing a politics of disguised protectionism and/or closure vis-a-vis many sectors of regional populations. These contradictions exacerbate difficulties in cultural understanding and can serve to confirm suspicions of neocolonialist or even imperialist designs.

Ambiguous external interpretations of EU geopolitics are not the only barrier to the realization of a wider Europe. The mere fact that the EU must expend considerable energy in maintaining a modicum of internal consensus indicates that, despite very ambitious objectives and visionary rhetoric, realpolitik will often favour pragmatic but suboptimal decisions. The EU has embraced the regional dimensions brought in by individual member states and groups of states as a means to forge greater consensus on its (as yet) comparatively sketchy foreign policy. But dimensionality in itself also creates rivalries having evolved out of specific regional perspectives of present and future EU countries, many of which can be considered to be 'on the periphery' of the Union. Contradictions in the many objectives pursued by the EU in engaging the member states of these regions appear inevitable as competition intensifies for the limited resources with which 'neighbourhood' programmes will be financed.

Discourses and processes of inclusion/exclusion can, in the long term, have a decisive impact on geopolitical outcomes and the stability of macroregional spaces. The evolution of a European political community within this context can therefore only be comprehended in terms of gradual institutional shifts, changes in attitudes and the ability not only to tolerate 'otherness' but also to break down barriers between 'East' and 'West'. Defining the borderlines between heterogeneity and coherence is not only the primary challenge of Wider Europe, it is also a central problem that will affect relationships between the EU and its larger 'Neighbourhood' for some time to come.

James Mittelman (1999) argues that, in scrutinizing its progressive and transformative qualities, regional cooperation should be analyzed with regard to the political opportunity spaces it creates for communities and different sectors of society to participate in economic and social development. As stated above, dimensionality and the notion of Wider Europe involves complex geopolitical strategies that are evolving into a policy initiative-cum-programme. Its main *raison d'être* is to provide orientation and coherence to the general goal of development and peaceful coexistence in a wider European context. It thus is a political script that reads very different indeed from the confrontational and/or antagonistic logic of militarized security. It is also a political agenda within which a wide variety of state and non-state actors have a role to play. If it is to succeed, it must, however, open up possibilities for a genuine transnationalization (rather than merely Europeanization) of space, extending networks, alliances and development opportunities to regions neighbouring the EU.

Chapter 3

Geopolitics of Scale and Cross-Border Cooperation in Eastern Europe: The Case of the Romanian-Ukrainian-Moldovan Borderlands

Gabriel Popescu

At the onset of the twenty-first century the eastern part of Europe is experiencing a series of multidimensional spatial changes that will have a lasting impact on the political, economic and social geography of the region. The EU enlargement of 2004 beyond the cold–war iron curtain to include eight former communist states (as well as Cyprus and Malta), together with the decision to admit two new East European members – Romania and Bulgaria – by 2007, represents an event of enormous significance for the European continent. It symbolizes the formal bridging of the east-west division of the continent and at the same time constitutes an opportunity for the building of a unified Europe based on shared values of peace and prosperity. However, the recent EU enlargement and its 2007 follow-up alone do not provide a panacea for the complex problems eastern Europe is facing. Not all former communist states in eastern Europe will become EU members by 2007, even though several of the remaining states aspire to membership. This context raises a set of concerns related to the incomplete unification of Europe. While the previous east-west dividing line through central Europe appears to have been bridged by the recent enlargement, it is possible that another dividing line is being established further east, at the future confines of the enlarged EU.

These changes in the political-territorial architecture of eastern Europe subject borders and border regions to greater scrutiny. Among the EU strategies for European unification, transboundary integration occupies a central role. Acknowledging the potential instability that might result from unsettled issues in East European border regions, the EU in conjunction with national governments in the region has made sustained efforts to liberalize border regimes. Over the past few years several Euroregions have been established in the Romanian-Ukrainian-Moldovan borderlands with the aim of offering an institutional framework for transboundary cooperation. Aware of the importance of such cooperation for the new European political-territorial situation, the Romanian, Ukrainian and Moldovan governments set aside their differences and supported the establishment of Euroregions in their border regions. However, unsettled disputes between these governments continue to cast their shadow on the process of transboundary cooperation in the region.

The EU has encouraged the creation of Euroregions along its future eastern borders because it saw in them devices that would promote stability and introduce East European countries to the process of European integration. However, the EU today indirectly hinders development in these Euroregions. In the light of potential EU membership of Romania by 2007 but without a comparable time frame for Ukraine and Moldova, the border between these three countries will become a EU border and remain so for some time to come. As Romania is implementing the EU *aquis communitaire* in preparation for EU accession, a stricter border regime is emerging at its eastern borders with Moldova and Ukraine. This situation has considerable impact on the momentum of cross-border cooperation in the Romanian-Ukrainian-Moldovan borderlands.

This study aims to contribute to an understanding of how multiscalar national, supra- and subnational forces interact to shape social life in these borderlands within the context of European integration. As the European Union's future enlargement will selectively include East European states, it is important to understand how the recently started process of cross-border cooperation, institutionalized in the form of Euroregions, will be affected by future EU membership or non-membership . Gaining insights into the complex relationships between the forces involved in the creation of new border regimes at the EU's future eastern borders can help shape policies and actions that will be more context-sensitive by taking into account these complex forces for the benefit of borderland inhabitants throughout eastern Europe.

For the present purpose, the supranational level will be represented by EU actions and policies, the national level by the relationships between the Romanian, Ukrainian and Moldovan national governments, and the subnational level by the activities of the concerned borderland inhabitants and their local institutions. Institutions at all three levels work together to produce outcomes in the Romanian-Ukrainian-Moldovan borderlands. For analytical reasons, the relationships between the supranational and the national levels, on the one hand, and the subnational, on the other, will be analyzed separately in the first part of the study. Then these levels will be reconsidered together in order to gain an understanding of the complex picture that emerges from their interaction in the borderlands of the three countries under study.

The Supranational and the Borderlands

The fall of communist governments throughout eastern Europe in 1989 opened new opportunities and challenges for the construction of a territorially integrated Europe. During the 1990s the EU was faced with a pool of postcommunist states desiring membership in the Union, but who suffered from an acute economic, political and social developmental gap when compared with the old member states. At the same time the region was rife with numerous simmering conflicts which had been frozen by the cold war and which had considerable potential to spread instability to the EU (Anderson and Bort 2001). This context was hardly conducive to addressing potential conflicts in the region; their resolution was therefore scheduled for a later date. Under these circumstances the desire of many East European states to obtain EU membership represented an opportunity for the EU to provide stability for eastern Europe by offering a framework for addressing various contentious issues in a cooperative manner. At the same time, EU leaders understood that if they engaged in sustained partnerships with East European countries aspiring to accession they would offer an opportunity for reducing the developmental gap between the EU and its neighbours, thus addressing some of the issues generating instability (ibid.).

The partnership between the EU and East European countries intensified after the 1997 Amsterdam summit when the EU decided to open membership negotiations with several of these countries. EU leaders conceived the integration of eastern Europe in terms of an eastward transfer of EU regional policies (Kennard 2003). The EU made cross-border cooperation one of the central elements of its transnational integration strategy – this was aimed at redefining centralized border-induced state territoriality to better address new challenges linked to East European borderlands.

EU-Driven Euroregion-Building in Eastern Europe

One way the EU envisions European integration is to promote cross-border cooperation in the form of Euroregions based on the principle of subsidiarity, that is on the idea that local authorities are better positioned than central governments to address the needs of their population (Loughlin 2001). Euroregions are territorially delineated regions situated on both sides of a state border, where close political cooperation and economic development are promoted for the benefit of the local civil society (Murphy 1996). Euroregions are thus considered institutional frameworks amenable to grassroots participation in cross-border territorial governance. This has the potential to generate bottom-up paths leading to cross-border integration that may ultimately succeed in transcending national borders and creating an integrated European space. To this end the EU has actively promoted and supported the establishment of Euroregions by encouraging national governments to support them and by setting up programmes to finance cross-border cooperation. Programmes designed to provide general assistance for eastern Europe, such as PHARE, for countries that had association agreements with the EU, and TACIS, for the CIS states, have been restructured to specifically assist cross-border cooperation by adding new subdivisions such as PHARE-CBC in 1994, PHARE-CREDO in 1996, and TACIS-CBC in 1996 (Ilies 2003).

The political, social and economic complexity of the Romanian-Ukrainian-Moldovan borderlands is representative of the nature of issues confronting borderlands in eastern Europe. The current borders between these three countries were established by the Allies at Yalta towards the end of the second world war and, for the most part, defined territories that had formerly belonged to Romania

(King 2000b). Throughout history these borderlands had repeatedly changed hands between Romania and the Russian Empire (and later the USSR) or between the Austro-Hungarian Empire (and, later, Czechoslovakia, Hungary and the USSR), in the western sector. The latest border changes took place after the breakup of the USSR in 1991. Territories that had belonged to the USSR became Ukrainian and Moldovan borderlands. Left unaddressed, the complex circumstances of these borderlands presented considerable potential for tensions in the region (Crowther 2000).

In this context, EU policies of cross-border cooperation appeared as one way to address potential issues in the region. Romania's aspiration to join the EU gradually made Romanian leaders receptive for these EU-inspired cooperation policies. Ukrainian and Moldovan leaders, too, understood that engaging in EU-endorsed cross-border cooperation with their western neighbour could result in closer relations with the EU that would bring them economic and political benefits (Prohnitsky 2002).

During the 1990s and early 2000s, four Euroregions were thus established along the Romanian-Ukranian-Moldovan border, even though there existed no comparable supranational framework such as the EU. Yet these Euroregions have the same goals, that is, to diminish the barrier function of state borders in order to enhance European integration. In the western sector of the Romanian-Ukrainian border, the Carpathian Euroregion, established in 1993 by Hungary, Poland, Slovakia, Ukraine and Romania, was the first Euroregion created exclusively between former communist countries (Ludvig 2003). Further to the east and south, Romania, Ukraine and Moldova established the Lower Danube Euroregion, in 1998, the Upper Prut Euroregion, in 2000, and Romania and Moldova the Siret-Prut-Nistru Euroregion, in 2002. The entire area formed by the Romanian-Ukranian-Moldovan borderlands are thus today covered by institutional structures of cross-border cooperation (Ilies 2004).

While these four Euroregions are technically national creations, they have benefited from assistance offered by non-governmental organizations (for example, the East-West Institute), supranational bodies (for example, the Council of Europe), transnational bodies (for example, the Association of European Border Regions) as well as others. In addition, the EU has provided direct help for cross-border cooperation, mainly in the form of financial assistance for concrete cross-border cooperation projects. While, during the early 1990s, the EU concentrated its financial aid predominantly on its immediately neighbouring borderlands (Poland, the Czech Republic etc.), a more comprehensive EU strategy including other East European countries has begun to take shape since the mid–1990s (Anderson and Bort 2001). Until then PHARE and TACIS national funds – the first for Romania, the second for Ukraine and Moldova – were only occasionally used to fund cross-border cooperation in these new Euroregions.[1] It was only in 2003, in preparation of

---

1   For a short period after 1994, the PHARE-CBC programme covered only the borders of Romania with other EU applicant countries such as Hungary and Bulgaria. Because Ukraine and Moldova did not have EU association treaties, the eastern borderlands of Romania did not

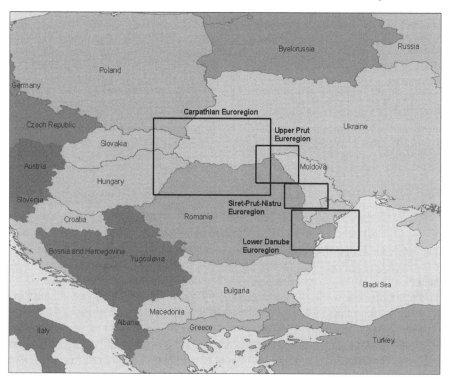

Map 3.1   Euroregions between Romania, Ukraine and Moldova

its eastward extension, that the EU expanded the spatial scope of PHARE-CBC to include the eastern borderlands of Romania with Moldova and Ukraine.

Although the amount of PHARE/TACIS funds allocated for cross-border cooperation in the Romanian-Ukrainian-Moldovan borderlands (several tens of millions of euros per year) has been rather modest when compared with other EU regional programs, these funds have nonetheless been of key significance for the functioning of such cooperation activities in the new Euroregions. Furthermore, EU funds brought into contact local authorities, NGOs and businesses from the borderlands, who then drew up funding proposals and implemented projects.

The EU enlargement of 2004 and the inclusion of Romania in 2007 will have direct consequences on cross-border cooperation in the borderlands under study here. Romanian membership in the EU will mean that a significant part of the eastern EU border and the border of Romania with Ukraine and Moldova will coincide for an approximate length of 1,330 km (approximately 650 km with Ukraine and 680 km with Moldova), thus creating new EU border regions further east.

---

benefit from PHARE-CBC funds until 2003, although their borderlands received TACIS-CBC funds since 1996.

It is in this context that the EU, in 2003, further redefined its strategy for cross-border cooperation by launching the European Neighbourhood Policy (ENP; see James Scott in this book). ENP is a comprehensive framework aimed at addressing challenges arising from economic, political and social disparities between the enlarged EU and its neighbours by fostering close partnerships with non-candidate countries, such as Ukraine and Moldova, in order to further stability, security and prosperity in the region (Lobjakas 2004). Within this framework, Neighbourhood programmes are already in place to support cross-border cooperation between Romania, Ukraine and Moldova. They will be followed, after 2007, by a New Neighbourhood Instrument that will provide increased direct financial support for cross-border and transnational cooperation along the external border of the enlarged EU and consolidate the various financial instruments available for cross-border cooperation in order to improve the management of these funds (Cilinca 2004).

The EU's Undermining of East European Cross-Border Cooperation

The new EU strategy towards East European non-member states embodied by ENP thus appears to be well suited for furthering cross-border integration in the Romanian-Ukrainian-Moldovan borderlands after the 2007 enlargement. However, a closer analysis reveals a series of contradictions embedded in the EU's new policies (Kennard 2003; Apap and Tchorbadjiyska 2004). Whereas privileged multilateral political, economic and cultural partnerships between the EU and East European non-member states will bring significant benefits to these countries, ENP falls short of offering Ukraine, Moldova and others the perspective of full EU membership. ENP is indeed being increasingly perceived as a substitute for EU membership for these countries (see James Scott in this book). Ukraine and Moldova, both aspiring to join the EU, hoped that the eastward shifting of the EU's borders would continue. However, the enlargement of 2004 seems to gave engulfed key EU members in 'extension fatigue', as the failed ratification of the European constitution by France and the Netherlands as well as the possible delay of Romania's accession by one year suggest. In this light, ENP appears to delay EU membership for Ukraine and Moldova rather than being a mechanism designed to achieve their rapid accession to the EU. One of the most worrying possible developments is the creation of a fortified EU external border in the east after the enlargement of 2007 because of EU concerns about 'soft' security threats, such as illegal immigration, organized criminality, and political and economic instability (Apap and Tchorbadjiyska 2004). Whereas an exclusive border regime at the external borders may address part of these security issues, its introduction will at the same time have negative consequences for the affected borderlands (Grabbe 2000). Perhaps the most important consequence is related to the application of the Schengen regime by Romania (Prohnitsky 2002). The Schengen regime regulates the circulation of persons, goods and ideas inside the EU as well as between the EU and non-EU countries. While its provisions liberalize cross-border flows of persons and goods inside the EU, they also reinforce the barrier function of the EU's external borders. In fact, the Schengen border regime is more nuanced, allowing a selective permeability of the EU borders by persons and goods deemed 'desirable' by the EU (Kennard 2004).

As part of its pre-accession strategy the EU, in 2002, extended to Romania certain Schengen provisions that allow Romanian citizens to travel to the EU without a visa. In exchange the EU required Romania to introduce stricter controls at its eastern borders, thus adding another layer of issues to cross-border cooperation in the Euroregions involving Romania, Ukraine and Moldova. As a part of the same pre-accession strategy the EU asked the Romanian government in 2004 to reinforce its eastern borders by introducing visas for Ukrainian and Moldovan citizens. The Romanian government successfully negotiated with the EU the delayed introduction of visas for Moldovan citizens, given the special relationship based on shared history between the two countries. However, travel visas were introduced for Ukrainian citizens in the summer of 2004.

In addition to visa-related issues, the adoption of the Schengen *aquis* by Romania has a wider impact on the region's borderlands. New customs regulations and import quotas for Ukrainian and Moldovan goods and services have been defined by the Romanian authorities in order to bring its border regime in line with EU standards. Furthermore, in order to meet EU security standards, the Romanian authorities are investing close to one billion dollars in border security (infrastructure) that might otherwise have been used to support transborder cooperation projects (Popa 2004).

The National Level and the Borderlands

The complex history of modern nation-state formation in Romania, Ukraine and Moldova left large ethnic minorities living in the borderlands of these three states, while abrupt territorial changes led to contested borders throughout the region. National borders between Romania and the former USSR were strictly enforced during the period that ended in 1989, and they have retained a heightened significance for the definition of national identity.

The emergence of Ukraine and Moldova, in 1991, as independent states within their former borders as Soviet republics significantly changed the geopolitical context of their borderlands. Both countries inherited many of the borderland issues that remained unsettled during Soviet times. The northern and southern sectors of the Romanian-Ukrainian border are composed of territories that Romania lost to the USSR with the 1939 Ribbentrop-Molotov Pact. The eastern part of the Romanian province of Moldavia thus became the Moldavian Soviet Socialist Republic (Bruchis 1994; Dima 2001). These territorial changes resulted in large Romanian ethnic minorities living in the border regions of the USSR and their presence constantly strained Soviet-Romanian relations. Moldova has a population of about four million, of which approximately 70 per cent are of Romanian descent (Biroul National de Statistica 2004). The population of Romanian descent in Ukraine as registered during the last Ukrainian census is divided between Romanians (151,000) and Moldovans

(258,000) (State Statistics Committee 2001). At the same time, there is a Ukrainian minority in Romania comprising around 61,000 inhabitants (National Institute for Statistics 2002).

In the early 1990s various Romanian political forces started to challenge Ukraine's rights to the borderlands lost to the USSR, generating tensions in Romanian-Ukrainian relations. However, the Romanian government did not openly embrace such claims, favouring instead a rather pragmatic approach in its relations with the Ukrainian government by promoting collaboration on issues of common interest considered to outweigh existing disagreements. After the mid–1990s, when Romania's perspectives of integration into Western cooperation structures such as NATO and the EU increased considerably, Romanian-Ukrainian relations improved as well, leading in 1997 to the signing of a bilateral treaty which confirms the existing land borders between the two countries and includes provisions for cross-border cooperation (Crowther 2000).

Romanian-Moldovan relations qualify as a 'special relationship' because of a given common history and cultural ties (King 2000a; Bruchis 1994). Romania was the first country to recognize the independence of Moldova in 1991, and it has since supported the cause of Moldova in the international arena. In the early 1990s there were expectations that the two countries would unite, following the German example. However, these hopes were short-lived and gave way to a 'silent integration' in which Moldova and Romania would continue to exist as separate states but maintain a privileged relationship leading to later unification. For the most part the two countries succeeded in maintaining such a relationship, translated inclusively into transborder cooperation, although in the early 2000s their bilateral relations reached a nadir (Prohnitsky 2002).

International Cooperation in the Romanian-Ukrainian-Moldovan Borderlands

The legacy of border problems and the lack of previous experience in cross-border cooperation influenced the perception of Euroregions in Romania, Ukraine and Moldova. In the early 1990s the national governments in the region regarded the establishment of Euroregions as a way to extend political control beyond their country's borders. Despite slow advances in trust-building between national governments, Euroregions were originally perceived as vehicles of 'incidentist' agendas and precursors of border revisions. They were therefore also viewed at the same time as direct threats to national integrity (Negut 1998).

Notwithstanding unsolved border issues between Romania, Ukraine and Moldova, there were real benefits to gain from cross-border cooperation in Euroregions. Such cooperation could in fact constitute a means to address the problems these countries faced even beyond the borderland dimension. In this context, Romania conceived Euroregions as mechanisms for maintaining institutionalized contacts with ethnic Romanians in Ukraine and Moldova. Ukraine, on the other hand, envisioned the

Euroregions established along its western borders as a gateway to EU integration, whereas for Moldova, sandwiched between Romania and Ukraine, Euroregions represented almost a necessity for its prospects of development.

Non-institutionalized forms of cross-border cooperation in the region appeared in the early 1990s, after Moldova's and Ukraine's independence. National governments rapidly liberalized the previously strict border regimes. Travel restrictions were lifted and the citizens of Romania, Ukraine and Moldova no longer needed a visa to travel to these neighbouring countries. Furthermore, in the context of the special relationship between Romania and Moldova, the citizens of the two countries did not need a passport to cross the border, but only a national identification card (Prohnitsky 2002). This situation represented an almost open border regime between Romania and Moldova that enormously boosted cross-border contacts at individual level, such as reestablishing family ties severed by over forty years of strict border regimes and promoting mixed marriages and cultural exchange, and generated small-scale development in the borderlands mainly through petty cross-border trade. At the same time the lax border regime also allowed the cross-border spread of criminal elements.

A breakthrough in institutionalized cross-border cooperation came in 1997, after the signing of a bilateral treaty between Romania and Ukraine. This treaty promoted the creation of Euroregions in order to assist ethnic minorities inhabiting border regions and to boost the economic status of these regions. The Romanian, Ukrainian and Moldovan presidents subsequently met in Izmail, a border town on the Ukrainian side of the Danube delta, to launch trilateral Romanian-Ukrainian-Moldovan cooperation within the region. Trilateral cooperation would represent a cornerstone of institutionalized cooperation in the borderlands. Although the Carpathian Euroregion was established as early as 1993, Romania did not fully participate in it until 1997. The launching of the 'Trilateral' was shortly followed by the establishment of the tripartite Lower Danube Euroregion in 1998, followed in turn by that of the Upper Prut Euroregion, in 2000, and of the Romanian-Moldovan Siret-Prut-Nistru Euroregion in 2002 (Ilies 2004). In the case of Moldova, these three Euroregions cover over 70 per cent of its territory and approximately 80 per cent of its population.

Within the framework of Euroregions certain progress was indeed made, particularly with regard to 'soft' sectors of cross-border cooperation. In 1999 a university branch was opened by Romania's Galati Lower Danube University in Cahul, Moldova, in the Lower Danube Euroregion (Ionescu 2000).[2] In addition, the Romanian government offered tens of thousands of scholarships to Moldovan and

---

2   For the most part academic personnel commuted across the border from Romania to teach on the Moldovan campus. In 2003, the number of students enrolled reached 400, most of them citizens of Moldova, and a smaller number from Romania and Ukraine. They were all supported by scholarships provided by the Romanian government. Beyond contributing to cross-border cooperation in the cultural realm, the university also represents an economic opportunity for the peripheral Moldovan borderland (Cranganu 2003).

Ukrainian students of Romanian descent studying at Romanian high schools and universities. These policies have had a strong impact on cross-border contacts in the borderlands of the regions since a significant part of these students came from Moldovan and Ukrainian borderlands and, for reasons of geographical proximity, often study in educational facilities located in the Romanian borderlands.

The Romanian government also granted Romanian citizenship, and thus Romanian passports, to Moldovan and Ukrainian citizens who lost their Romanian citizenship after their country's incorporation into the USSR or who can prove Romanian ancestry. It is estimated that by 2000 more than 300,000 Moldovans and Ukrainians had acquired Romanian passports.[3] Since 2002, when Romanian citizens were allowed to travel to the EU without a visa, the possession of a Romanian passport by Moldovan and Ukrainian citizens has become highly desirable and will be even more so once Romania will have joined the EU and adopted a EU visa regime for citizens of East European non-member states.[4] The primary motivation for large numbers of Moldovan and Ukrainian citizens to acquire Romanian passports is economic. For instance, the number of Moldovan citizens who have left their country since the prolonged economic downturn starting in the late 1990s and who are now working temporarily – and most often illegally – in various European countries is estimated at over 600,000, or 25 per cent of the population (Jandl 2003). For those who have remained in the borderlands, the possession of a Romanian passport offers a number of opportunities for cross-border petty trade, in addition to guaranteeing easy border crossing into Romania and into the EU.

To sum up, international cooperation between Romania, Ukraine and Moldova during the 1990s and early 2000s led to the liberalization of the border regime and eventually to institutionalized cross-border cooperation in Euroregions. While these actions did not succeed in generating an economic miracle in the borderlands and rarely led to a real devolution of power to local authorities, their significance resides in increasing the scope and the level of local cross-border interaction and in creating a framework for national governments to address contentious borderland issues.

Borderland conflicts between Romania and Ukraine

In a context where institutionalized cooperation in Euroregions remains to a large extent a top-down enterprise, disputes between states, whether involving borderlands directly or not, are likely to have an impact on their border regions. One such contentious issue between Romania and Ukraine is related to the delineation of the maritime boundary and the continental shelf around Serpent Island, a tiny barren, uninhabited island in the Black Sea, inherited by Ukraine in 1991 and situated at

---

3   *Adevarul*, 11 November 2002.

4   The Romanian media have reported that organized crime networks have already tried to take advantage of Romanian citizenship laws. In Ukraine, for example, the price for the procurement of false documents to attest Romanian descent can be as high as € 1,700 *(Adevarul*, 11 November 2002).

approximately 30 miles from the Romanian city of Sulina (Cucu and Vlasceanu 1991). By virtue of the Paris Peace Treaty (1947) the island belonged to Romania which ceded it to the USSR in 1948, when the latter built a military outpost to control navigation at the mouths of the Danube River. However, a protocol signed for this purpose has never been ratified by the Romanian parliament. Nor did Romania and the Soviet Union ever successfully negotiate the delimitation of their maritime boundary. Thus, from a legal point of view, the island has remained Romanian territory.

In the early 1990s the Romanian government reclaimed the island, first from the USSR and later from Ukraine. Romanian-Ukrainian negotiations regarding the delineation of their maritime borders started in 1995 as part of the negotiations for a bilateral basic treaty between the two countries but an agreement could not be reached. Yet the basic treaty was adopted in 1997 with the stipulation that negotiations between the two governments regarding the delineation of the continental shelf would continue and, if no agreement could be reached, that the two countries would appeal for arbitration to the International Court of Justice. In the meantime NATO and EU membership required Romania to settle its territorial claims. Instead of claiming repossession Romania therefore demanded the delineation of ordinary maritime borders on the basis of Romanian and Ukrainian shorelines. However, even as major progress in Romanian-Ukrainian negotiations was being reported, the dispute flared up anew in the early 2000s after the announcement that significant oil and gas deposits had been discovered in the continental shelf around Serpent Island. Romania, stressing the uninhabited character of the island, considers the island a rock, which would result in the establishment of an exclusive economic zone (EEZ) smaller than the standard 12 nautical miles surrounding the island and therefore represent an opportunity for Romania to share some of the oil and gas deposits. Ukraine for its part has set out to prove that Serpent Island can be inhabited and would thus qualify for island status. For this purpose it has maintained a military presence on the island.[5] The inability of the two governments to reach an agreement on sharing access to the continental shelf around Serpent Island has caused the issue to be brought to the attention of the International Court of Justice, and oil companies are presently holding off their drilling operations until the diplomatic conflict will have been settled.

Another contentious issue in Romanian-Ukrainian relations is the work Ukraine started in 2004 on the construction of a canal to secure an independent connection between the Danube and the Black Sea (Anderson 2004). The southern sector of the

---

5   Some 50 soldiers and their families are regularly shuttled to and from the island, and water and other resources have to be brought in from the mainland by helicopter. In its attempt to convince the international community that Serpent Island qualifies as an island, the Ukrainian government intends to demilitarize it and create a series of meaningful economic activities there. A phone line has been installed, military barracks have been modernized, and fishing infrastructure is being developed. In 2004 a Ukrainian bank announced that it had opened a branch on the island, although there was no meaningful commercial activity there.

Romanian-Ukrainian border, which is part of the Lower Danube Euroregion, passes through the Danube delta.[6] The border runs along the Chilia channel, the northernmost of the three main Danube channels that form the delta. The other two main channels, Sulina in the centre and Saint George in the south, belong to Romania, together with most of the delta. International conventions that regulate navigation on the Danube have designated Sulina as the international shipping route to the Black Sea (Anderson 2004). The Chilia channel is the largest of the Danube channels, transporting almost 60 per cent of the water. It also carries a high volume of sediments that are deposited at the mouth of the channel, forming a secondary delta on Ukrainian territory and making navigation difficult for high tonnage ships. Ukrainian authorities therefore decided to use the Bystroe canal, an offshoot of the main Chilia channel, to build a waterway between the Danube and the Black Sea on Ukrainian territory. However, the Bystroe canal partly runs through a UNESCO reservation, where major human activities are prohibited. Romania vehemently opposed the construction of the Bystroe canal on the grounds that it would produce an ecological catastrophe in the whole Danube delta, well beyond Ukrainian territory, whereas Ukraine argues that the construction of the canal will bring economic benefits to peripheral Ukrainian borderlands along the Chilia.

The issue of the Bystroe canal turned into an intergovernmental dispute and mobilized civil society in Romania and Ukraine, as well as in other European countries. Non-governmental organizations in various European countries, the European Commission, the Council of Europe, the US State Department and UNESCO have taken up the issue asking Ukrainian officials to stop the project until its ecological consequences will have been studied. However, both Romanian and EU negotiations with Ukraine have been unsuccessful, and the construction of the canal has continued. The Romanian government eventually brought the issue to the attention of the International Court of Justice. The Romanian authorities also announced that they would study the possibility of building a canal of their own that would take water from the Chilia channel and render the Bystroe canal useless (Dutu 2004).

The significance of the Bystroe canal seems to extend beyond economic and ecological issues. As Romania is now a NATO member and is scheduled to join the EU in 2007, an independent waterway connecting the Danube to the Black Sea becomes an important strategic asset for Ukraine. Furthermore, in light of the newly discovered oil and gas reserves around Serpent Island, an independent shipping route for the oil extracted there would undoubtedly benefit certain Ukrainian interest groups. However, it is less clear for now to what extent borderland inhabitants and their traditional way of life would benefit from the construction of the canal.

---

6   This delta is the largest in Europe and has a unique ecosystem that has received the status of an UNESCO reservation.

Conflicts between Romania and Moldova

After the 2001 elections, when the Communist party came to power in Moldova, the special Romanian-Moldovan relationship abruptly deteriorated. Later that year bilateral relations reached an all-time low when in a speech addressed to the European Court of Human Rights (ECHR) in Strasbourg the Moldovan minister of justice accused the Romanian government of 'expansionist policies' in Moldova (Constitutional Watch 2002). The allegations were made after the ECHR ruled in favour of the Romanian-language Bessarabian Metropolitanate in its case against the Moldovan government. During Soviet times the Orthodox Church in Moldova came under the jurisdiction of the Patriarch of Moscow and All Russia. After Moldovan independence, the Metropolitanate of Chisinau and All Moldova was organized under Russian jurisdiction to assert control over religious matters in Moldova. The Bessarabian Metropolitanate, which had managed religious matters when Moldova was still part of Romania, unsuccessfully demanded official recognition as the representative of the Romanian-speaking Moldovans and, in 1998, took the case to the ECHR. The issue of the Bessarabian Metropolitanate is part of a broader set of issues straining relations between the two countries and affecting the state of cross-border cooperation between them. Moldovan political life is dominated by two main currents of opinion divided between left-wing pro-Russian and right-wing pro-Romanian forces. As a result, Moldovan international politics since independence have been vacillating between closer ties with either Romania or Russia. Since 2004 Romanian-Moldovan relations have, however, improved again.[7] Facing general elections in early 2005 and fearing a repeat of the events in neighbouring Ukraine that brought the opposition to power, the Moldovan communist government appears to have experienced a change of heart in international politics. It now seems to collaborate more closely with Romanian and EU officials since the launching of ENP by the EU, of which Moldova is a direct beneficiary.[8]

Negative Outcomes of Multiscalar Policies in the Romanian, Ukrainian and Moldovan Borderlands

It is difficult to separate supranational EU policies from national policies when examining their outcomes for the borderlands, as these policies often work together. While the EU, for instance, is responsible for asking Romania to introduce visas for Ukrainian and Moldovan citizens, national governments are negotiating the particulars of the visa regime. Conversely, in the case of international disputes, national governments are establishing policies affecting the borderlands and subsequently appeal to international organizations to back them up. However, local institutions, though generally less powerful, should not be seen as passive. They are part of multiscalar politics and are reacting to national and supranational policies

---

7   Radio Free Europe/Radio Liberty, 21 January 2005.
8   Radio Free Europe/Radio Liberty, 23 February 2005.

affecting them. This multilevel interaction generates a complex political-territorial environment in which cross-border cooperation has to operate.

At the national level the conflicts between Romania, Ukraine and Moldova have had a chilling effect on cross-border cooperation in the three Euroregions. After the worsening of Romanian-Moldovan relations in 2001, the Moldovan government attempted to close the Lower Danube University branch in Cahul, invoking the lack of a bilateral protocol in the field of education between Moldova and Romania. This decision is opposed by borderland inhabitants who perceive the university as a source of cultural and economic regional development. Moreover, the demand for scholarships in Romanian educational establishments continues to increase. The Romanian government also introduced import restrictions for certain Moldovan goods, justifying its decision by invoking EU requirements. However, it appears that the EU had not asked Romania to introduce these restrictions without delay.

Similarly border disputes between Romania and Ukraine have had a direct impact on their borderlands. The dispute about the continental shelf and the building of the Bystroe canal antagonized many borderlands inhabitants who found themselves in two opposing national camps. When members of a Romanian NGO based in the borderland city of Galati, Romania, went on a fact-finding field trip to the Ukrainian border town of Chilia, they were received by local inhabitants protesting Romania's opposition to the building of the canal (Moisi 2004).

More recently requirements of a visa for Ukrainian and Moldovan citizens travelling to Romania, as well as trade-related restrictions, have been of utmost concern to the inhabitants of the borderlands (Gheorghiu et al. 2002; Malynovska,2002). There is a widespread feeling of exclusion among the inhabitants of the Ukrainian and Moldovan borderlands (Apap and Tchorbadjiyska 2004).[9] One reaction is to pursue their quest for acquiring Romanian citizenship and a Romanian passport in order to breach the 'wall' (Lippert 2001). The Moldovan economy remains indeed largely dependent on remittances sent home by over one-third of its population, and the possession of a Romanian passport allowing free travel to Romania and to the EU has become a crucial asset.

The tightening of the border regime at Romania's eastern borders also reinforces social inequalities in the borderlands and could lead to a widening of the development gap between the EU and its eastern neighbours (Lippert 2001). It is often assumed that the impact of the Schengen regime on the borderlands will be diminished to a certain extent by the fact that a large number of inhabitants of the Moldovan and Ukrainian borderlands already hold a Romanian passport. This is, however, not the case. Many people involved in cross-border petty trade between Moldova and Romania are poor Moldovan farmers living near the border (Gheorghiu et al. 2002). Often, this is their

---

9   This feeling also has a visible physical dimension that adds to the already established psychological dimension. While travelling in 2004 in these borderlands, I saw that a series of newly-built Romanian border checkpoints were quite prominent in the landscape, while on the Moldovan or the Ukrainian side passport checking takes place in more modest settings, or even in military-like barracks.

only means of subsistence. Their income is so low that many are unable to afford a passport, let alone the official fee – several hundred dollars – required to apply for Romanian citizenship. Their livelihood was affected for the first time in 2001 when Romania introduced passport controls at the border checkpoints with Moldova.[10] The introduction of costly Schengen visas after 2007 will worsen these farmers' situation, at least in the short term. As a result an increased number of Moldovan and Ukrainian citizens have decided to establish permanent residence in Romania, particularly in the border areas where many have relatives. However, during the past few years the Romanian authorities have dramatically slowed the process of granting Romanian citizenship at the request of EU officials concerned by liberal Romanian citizenship laws (Lippert 2001).[11] The prospects of a strictly enforced EU border regime seems to lead to an increase in illegal and semi-illegal activities, such as smuggling of persons and drugs, and in corruption. Large-scale smuggling activities are most probable to move their focus from the Romanian-Hungarian border region, where the current EU border is located, to the Romanian, Ukrainian and Moldovan borderlands.

Imposing a strict border regime at the EU's external borders will also result in reinforcing the central governments' grip over the borderlands, which in turn could reinforce their peripheral status. The recent experience of cross-border cooperation in Euroregions has been fostering a certain degree of initiative on the part of the local authorities and, to a lesser extent, of the local small businesses. Overemphasizing the security dimension of borders will lead to an increased direct and indirect presence of central governments in the borderlands. This state of affairs can reverse the modest progress made toward administrative decentralization and can relegate border regions to the peripheral status that traditionally defined them in eastern Europe.

The border regime between Romania and Ukraine and Moldova has already undergone considerable tightening in the last few years with the reintroduction of passport controls at the Romanian-Moldovan border in 2001 and the introduction of visas for Ukrainian citizens in 2004. These measures have already resulted in a decrease of the Euroregions's activities in general, and of `people-to-people' contacts in particular. The Schengen regime is thus likely to undermine the role of Euroregions as promoters of economic development and political cooperation in East European borderlands. Reinforcing the external borders will not solve problems of illegal immigration and criminality but shift them to the borderlands. The EU will therefore have to find other ways to address these issues, with ENP as a useful

---

10 Being aware of this situation, the Romanian government granted one million dollars to the Moldovan government to be used specifically for issuing Moldovan passports to poor farmers who inhabit the borderland (Chomette 2002).

11 As the possession of a Romanian passport offers the possibility of working in Romania while residing at least temporarily in Moldova or Ukraine, the curtailing of this economic and social avenue contributes to the decision made by many people, such as students who finish their studies in Romanian institutions or mixed couples, to establish permanent residence in Romania in order to avoid the complex Schengen visa application process they would have to undergo were they to reside in Moldova or Ukraine.

starting point. There is clearly a need for EU policies to reach the less privileged borderland population through a special visa regime and facilities for small cross-border trade.

Conclusions

Multidimensional political, economic and social changes that took place in eastern Europe in the early 1990s had a strong impact on the geopolitical architecture of the European continent: previously 'frozen' conflicts flared up in eastern Europe, and new national borders appeared in the region. The EU took the opportunity to expand eastward in order to address potential sources of instability coming from the region and to unify the continent. In preparing East European countries for European integration, the EU adopted a strategy based on cross-border cooperation institutionalized in Euroregions and aimed at a gradual lessening of the barrier function of national borders and at integrating previously divided border regions in order to build an integrated European space. Since many national governments in the region desired EU membership, they became increasingly receptive to this strategy, despite continuing divergences. By 2002 four Euroregions were established across the borders of Romania, Ukraine and Moldova. Additionally, these three countries have taken important steps since the early 1990s to liberalize their previously strictly enforced border regimes. Despite difficulties encountered by Euroregions due to intergovernmental conflicts, cross-border cooperation between Romania, Ukraine and Moldova gained momentum. Contacts across the border between local authorities took place regularly, and interpersonal contacts among borderland inhabitants soared during the 1990s.

In 2004 the EU expanded eastward by including eight East European countries. In addition, two more countries – Romania and Bulgaria – are scheduled for inclusion in 2007. Although this EU expansion reached historic proportions, it also generated further challenges for those East European countries that do not have immediate prospects for EU membership. In particular, there is considerable concern that the EU's eastward expansion will pause for a long period and that the EU's new external borders will come to restrict the circulation of people and goods. Signs of a stricter border regime are already present at Romania's borders with Ukraine and Moldova, despite the fact that Romania is not yet a member state. The previously liberal border regime has been continuously tightened at the EU's request since the early 2000s. After 2007 the citizens of both Moldova and Ukraine will need Schengen visas to travel to Romania. It remains uncertain whether ENP (and in particular its financial support for cross-border cooperation) will provide the necessary framework for creating the dense interdependencies of successful cross-border cooperation. Present policies, whether designed by the EU or by national governments, show little consideration for the interests of border regions. It appears that the degree of support for cross-border cooperation in Euroregions at both supranational and national levels is contingent on interests agents at these levels share with borderland inhabitants.

To date the latter are not appropriately considered in the design of supranational and national policies that affect them. For the most part, inhabitants of borderlands react to these scalar policies by trying to navigate through them in the hope that their reactions ultimately will strain these policies and thus force their modification.

Chapter 4

The European Community as a Gated Community: Between Security and Selective Access

Henk van Houtum and Roos Pijpers

Living is easy with eyes closed
Misunderstanding all you see
Nothing is real and nothing to get hung about
Strawberry Fields forever
— The Beatles, 1967

Such a pretty house
Such a pretty garden
No alarms and no surprises
Please
— Radiohead, 1997

Fear of (mass) migration has been and still is an important aspect of ongoing processes of sociospatial bordering of immigrants within the European Union (EU) and its various member states. The recent EU enlargement involving many post-Soviet nations has but intensified these sentiments. Offering an alternative to the well-known yet flawed Fortress Europe metaphor, the authors of this contribution argue that the moral panic caused by immigration and the migration policies adopted by (various member states of) the European Union follows a geostrategic logic which closely resembles that at work in the management of a gated community.

Ambivalent Immigration Policies

Over time, but especially since the creation of an internal market, the EU has 'modernized' its immigration policy, with a specific focus on containing asylum migration, fighting irregular/illegal migration and extending European migration policy to countries of origin and transit. Non-EU countries are urged to more firmly control emigration, and development aid is increasingly tied to agreements obliging so-called third countries 'to take back illegal migrants'. Furthermore, all non-EU countries on the edges of Europe receive both political and financial incentives to reinforce their border controls. Liberalization of cross-border labour mobility and moral equality for 'all' EU citizens living in countries that are part of the internal market are combined with interim measures restricting the freedom of movement for the Union's new citizens and efforts to tighten control at the new external borders. This new regime means that the EU's borders have been increasingly closed, fortified and policed to the point where attempts to escape detection by border guards can end in death for many would-be immigrants (see Harris 2002). Indeed, the life of refugees who are trying to find work or shelter in the European Union is being criminalized by these border policies; perversely, their deaths are implicitly seen as 'collateral damage' in the struggle against illegal migrants. Estimates of deaths at the borders differ, but many would agree that it is presently in the thousands. At the same time, however, and in sharp contrast to this policy of border closure, strategically selective political measures are being implemented in order to attract labour from outside the EU to help bypass shortages in specific economic sectors. It is this little investigated policy contrast that we wish to clarify and analyze in more detail.

The Desire for Border Management

We argue that 'desire' is a key notion with which to capture the contradictions of the EU's border and immigration policies. On the one hand, we are witnessing an unrelenting desire for sovereign and autonomous control of the movements of those who are seen as 'redundant fortune seekers'. Here 'desire' expresses itself as a wish to control the numbers of the 'redundant' and the mobility of persons perceived as 'non-western' immigrants or refugees in order to preserve social cohesion within the member states of the EU. On the other hand, there is also a growing – and more and more openly voiced – desire to selectively invite non-EU citizens to help safeguard the EU's long-term economic welfare in the face of demographic, knowledge-based and skill-specific shortfalls in European labour markets (van Houtum 2003). Although these two forms of desire are intrinsically contradictory and very difficult to sustain, let alone manage, in combination, policies establishing either 'desirability' and 'undesirability' of immigration are a means to the same end, that is to protect the EU's own internal 'comfort zone' (ibid.).

In general, the desire to regulate borders of opening and closure could be considered as the inverse of fear. This 'fear' has many faces. Sometimes it shows itself as a fear of the Other, of the unknown and of the 'stranger' and is associated with the fear of losing a self-defined social identity and the meticulously constructed feeling of fitting within a certain territorial community. At other times it is related to material concerns such as job losses and/or decreasing national welfare and social benefits. Despite its many forms, fear of immigrants is rarely grounded in a thorough knowledge of current EU realities, that is the EU's lagging market competitiveness vis-a-vis more efficient countries such as the US and Japan and cheaper producers such as China and India, or of the enlargement process and the geoeconomic transformation of new member states. EuroBarometer reports, for example, indicate that both advocates and opponents of enlargement share the opinion that accession would negatively affect their own domestic employment situation,[1] while a majority of respondents were unable to name even three applicant member states. Since May 2004, it has become evident that none of the fears voiced over 'downward convergence' were justified. And yet lack of information about EU realities continues to inform current migration discourses dominated by dissatisfaction with past immigration and minority integration policies, thus leading to the reproduction and sustenance of the dichotomy between 'us, Europeans' and 'them, non-Europeans' (van Houtum 2003). Illustrative of this is the fact that almost all 'old' member states of the EU (with the notable exception of Ireland, the United Kingdom and Sweden) have imposed so-called waiting periods or delaying measures restricting access to the labour market for citizens from new EU member states, such as Poland, Hungary and the Czech Republic, for up to seven years. Incorporation of the new EU members into the Schengen zone will therefore take place in stages, after the successful setting-up of a *cordon sanitaire* around 'core Europe'.

At the same time that 'buffer zone' politics have been established within the EU, new forms of border management have come into existence not only at border locations proper, but also at internal locations such as airports and asylum centres – a telling example are practices at the highly controversial 'detention centres' recently installed in the Netherlands, which are destined to efficiently conduct deportation procedures for rejected asylum seekers and undocumented migrant workers, many from countries with association status, such as Romania, Bulgaria and Ukraine.

Moral Panic in the EU

The fear of massive westward flows of migrants after EU enlargement, itself embedded in a broader context of 'negative' migration discourses, demonstrates that, within society, there exists a continuous desire for opposing the normal to the deviant and the self to the Other. The Other is utilized to compare, to associate with and/or to oppose to (Sibley 2001; Derrida 1973; Luhmann 1985). Without Others there is no need to have a social identity (Jenkins, 1996), for it is only in the awareness of imaginative Others that identity becomes a relevant and contingent source of meaning and experience. This negative conception of social identity resonates with Bauman's (1997) argument that 'each order has its disorder and each purity its own dirties' and Sibley's (1995, 2001) well-known notion of the 'purification of space'. By definition therefore, a border deconstructs a difference (the outside in and/or the inside out) but at the same time creates a difference (a new outside). Yet the function of b/ordering is precisely that – the making of a divisive order in an assumed chaos.

---

1 EuroBarometer public opinion surveys are conducted each spring and autumn by the European Commission and consist of identical sets of questions submitted to representative samples of the population aged 15 years and over in each member state. The November 2002 survey shows that no less than 31 per cent of respondents in favour of enlargement expected unemployment numbers to worsen after EU accession, a view shared by 61 per cent of respondents who opposed enlargement.

In this way, what lies beyond the border can be justifiably neglected or treated with indifference (van Houtum and van Naerssen 2002). A spatial imaginative bordering process accordingly rests upon the redefinition of 'friends' as 'natives' (Bauman 1990), among whom common assets of knowledge and wealth are constructed and distributed (Giddens 1984). On the other hand, 'strangers' are granted residential rights only if this is seen as desirable, and this despite that fact that the desirability is often disguised as 'feasibility' (Bauman 1990). The identity of strangers is therefore usually not a matter of choice (Bradley 1997; Miller 1995). Strangers are excluded on the basis of their different (or absent) nationality as manifested by country of birth, colour, creed or culture. Strangers must adjust to the identities and nationalities of their host society if they wish to be included. Each society then, as Bauman famously argued, 'produces its own kind of strangers' (Bauman 1997, p.17).

Depending on the circumstances in individual EU member states, the desire for b/ordering has found new sociopolitical outlets and performance spaces, thereby often creating a new normative vocabulary. Under present conditions it could be argued that the dominant (and even disciplining) discourses on the need to 'communalize', as expressed in terms such as 'common market', 'internal market', 'a borderless Europe' and 'the need for European citizenship', have invoked a notion of 'abnormality' represented by people living outside the EU and by refugees from non-EU countries seeking shelter inside the Union. The result is increased anxiety and fear of the Other, or in the words of Sibley (2001), 'moral panic'. As a result of moral panic, communities seek to eliminate contested spaces and liminal zones by appropriating such spaces and excluding the offending 'Other' (Sibley 1995, p. 39). As attempts are being made to create such a categorical difference between citizens of EU and non-EU countries without abandoning what is seen as politically correct, there has been a constant search for appropriate definitions and terms for 'non-insiders', that is, people from outside the EU. Indeed, many terms are used, such as strangers, aliens, foreigners, newcomers, (im)migrants, refugees and the Dutch term *allochtonen*,[2] to name but a few. By defining a border between normality and deviance, the defining, making and exclusion of the Other is, as Sibley terms it, a 'colonization' of social life. This colonization is a way of creating difference (through the use of social space) beyond boundaries and of rejecting difference within bounded spaces.

Fortress Europe?

There is no shortage of normative positions on the EU's exclusion of the Other. Walzer (1983), for instance, provocatively states that communities should not be allowed to claim territorial jurisdiction and thus legitimacy to rule over all people with whom they share a territory. He argues that, although admission and exclusion are at the core of community independence, the rule of citizens over non-citizens and

---

2   The Dutch term *allochtonen* refers to people born 'elsewhere' and literally means 'of different soil'.

members over strangers is 'an act of tyranny'. Seyla Benhabib, following Derrida's essay on hospitality in her plea for 'interactive universalism', agrees with Walzer on this point (Derrida 1973; Benhabib 1996). For, she asks, what is the ethical difference between the right to leave a democratic country, since in democratic societies citizens are not prisoners, and the right for others to enter? Following this line of reasoning, inhabitants of the 'chaos' outside the insulating boundaries of union are the new undesired 'barbarians' and hence denied access. It is no wonder then that the European Union can resemble a fortress where unwelcoming and even hostile visualizations of closure abound, or a maze or sieve to those not intimidated by the EU's efforts to fortify its external borders (see also Kramsch et al. 2004).

However, we would argue that the Fortress Europe metaphor, besides its all too dramatic ring and its being a geographical misnomer (it is the European Union, not Europe), is also increasingly untenable. For while the interception and the frequent death of 'illegals' at the gates of the EU fit the image of a fortress, they sharply contrast with policies encouraging the immigration of individuals representing economically desirable and scarce forms of labour. Since many nations of (western) Europe are trying to cope with shortages of specific (academic) knowledge or skills, economic demand for foreign experts is often made explicit in work and residence permits and visas granted to non-EU migrants. For this reason politicians in several countries have started a debate on whether to introduce Green Cards (Germany), work permits (Great Britain), quotas (Italy) or a speed-office (The Netherlands) enabling desired foreign employees to bypass bureaucratic immigration procedures. Top managers, engineers, PhD students and talented soccer players from third countries all can be strategically selected for immigration by non-state actors such as large firms, universities and specialized employment agencies. In the case of the new EU member states, workers with medium-level and, to a lesser extent, low-level skills (for example, nurses and seasonal workers in agriculture or construction) may occasionally be granted access to the EU's labour market, too.

As a result, the buffer zone politics described above do not apply to this limited group of east-west migrants. In contrast with the 'anti-redundancy' and 'anti-burden' politics being applied to the majority of non-EU foreigners, a few of them are seen as valuable 'assets' and are most welcome in the European internal market. Moreover, a growing body of research produced by supranational organizations suggests that a softening of immigration and border policies will be unavoidable in present EU member states as the absolute decline in the active workforce will continue into the next decades (see also Sassen 2002). In its 2000 communication On a Community Immigration Policy, the European Commission acknowledged the need for 'replacement immigration' in the near future, strengthening again its ambivalent views on migration (Commission of the European Communities 2000). Although recognized by individual member states, the unavoidable need for more immigration – if sustainability of the EU's market competitiveness is to be considered at all important – does not appear to have an impact on the short-sighted and restrictive migration policies now in place.

Such strategically selective and exclusionary sociospatial bordering processes of migration can be understood as one result of the globalized capitalist economy. The European economic space and its labour markets, 'fortified' or not, have not been able to escape the global reach of economic agents and processes. In one of her earlier writings on the subject Saskia Sassen (1988) comments:

> National boundaries do not act as barriers so much as mechanisms reproducing the system through the international division of labour. [...] The enforcement of national borders contributes to the existence of a large number of countries in the form of a periphery and the designation of its workers as a labour reserve for global capital. Border enforcement is a mechanism facilitating the extraction of cheap labour by assigning criminal status to a segment of the working class – illegal immigrants. Foreign workers undermine a nation's working class when the state renders foreigners socially and politically powerless. At the same time, border enforcement meets the demands of organized labour in the labour-receiving country insofar as it presumes to protect native workers. Yet selective enforcement of policies can circumvent general border policies and protect the interests of economic sectors relying on immigrant labour (p. 36–7).

There exists of course a wide range of interpretations and assessments with regard to these developments. The general neoliberal approach, which with its commitment to free trade, privatization and deregulation has become the dominant political paradigm of our time, is seen by many as having a negative impact. The term 'neoliberalism' has become almost synonymous with exploitation, including the exploitation of labour (see, for instance, Sparke 2002). Favell and Hansen (2002), on the other hand, provocatively argue that market-driven strategic selectivity is irrevocably becoming a major determinant of migration flows in an enlarging European Union. In their view, the 'normative' Fortress Europe is quite open in 'positivist' reality, both for economic migrants (through strategic selectivity) and asylum seekers (through highly inefficient and ambivalent asylum procedures). The idea of Europe would, accordingly, benefit greatly from drawing a much clearer line between economic migration and asylum.

A similar, yet more fundamental assessment is provided by Slavoj Zizek (1998), who, through his idea of 'post-politics', calls attention to the de-politicization of European politics as he sees it (see also van Houtum and van Naerssen 2002). Post-politics goes beyond ideological differences between left and right in order to explain the irreversible expansion of global capitalism as consensus politics, wherein universal social issues such as transnational migration are reduced to mere administrative procedures (Zizek 1998 and the interview with him in Deichmann, Reul and Zizek 2000). From being an initiator of radical changes in the world order, politics have turned into an end-user of manoeuvring space within existing parameters. In this respect, the Comaroffs go as far as to argue that by hiding its own ideological underpinnings in the dictates of economic efficiency and efficacy, politics portend its own death (Comaroff and Comaroff 2002).

The European Union as a Gated Community

What then is left of the Fortress Europe rhetoric once selective access of desirable immigrants has been taken into account? The border is economically closed for an overwhelming majority, yet open for some. To understand and better grasp these paradoxical border policies we have to ask ourselves what exactly it is that we are trying to protect in the EU? In economic terms, protection principally concerns comfort, which itself is an (economic) interpretation and extension of the concept of 'easiness'.[3] Thus, the interpretation of the border in economic terms relates to the degree of protection of the national economy, of 'our national interest'. The latter is about appropriating and justifying comfort. A territorial border, from an economic point of view, is therefore a form of territorialization of wealth and potential. The chances for strangers to be allowed to play a role in the national economic arena are higher when estimated national wealth and employment effects brought about by their admission are overall positive.

In its rather opportunistic search of inputs in order to maximize competitive advantage, the EU actively seals off specific domains of the economy for and at the expense of others. Probably the best-known example of this is the agricultural sector, which, in most of the member states, is unable to compete with efficient large-scale production elsewhere. Instead of letting 'the market' prevail here, as neoclassical ideology would dictate, highly protectionist policies (import tariffs, export subsidies) prevent non-EU agricultural goods from penetrating the European Union's economic space. This is quite similar, we would suggest, to current migration policies in the enlarging European Union that protect the comfort of job security in a highly inflexible labour market and in a number of highly uncompetitive economic sectors. By denying outsiders access to domestic labour markets, the EU protects its domestic workers through border enforcement and by issuing (expensive) work and residence permits. The sustenance of comfort, the amount of money and/or (the growth of) wealth gained is thus a form of collective self-interest of the community of human beings who call each other 'members' of the EU.

Hence we would argue that, much more than a fortress, the European Union, through its protective immigration policy, is beginning to look like a 'gated community'. A gated community, also referred to as a 'defended neighbourhood', is a form of housing found mainly in developing countries with large internal income differences such as Mexico, Brazil and Venezuela, as well as in the United States. In such communities the affluent gate themselves off from the rest of society in an enclave, primarily out of fear and a desire to protect their property and well-being. In an excellent recently published empirical overview, Blandy et al. (2003) provide the following definition of gated communities: 'walled or fenced housing developments to which public access is restricted, often guarded using closed-circuit television and/or security personnel, and usually characterized by legal agreements (tenancy or leasehold) which tie the residents to a common code of conduct' (p.2).

---

3   The meaning of the Latin *comfortare* is 'to strengthen', 'to ease'.

Gated communities express a clear-cut form of lacking sociospatial solidarity and of purification of space by shutting their gates to the 'outside' world under the banner of privacy, comfort and security. Non-members, usually the non-rich, are excluded from these spatially bordered contractual associations. A gated community is a kind of frontier land that is predominantly built and maintained by the private sector. Membership is paid for and non-members are labelled guests. The gates of such a community are not only a result of the desire to produce a specific space for the outsider, the stranger, but even more so a purified space for the insider. It is the commercialization of fear. It does not come as a surprise then that the identity of community members is marketed as a life style, as a status that you buy. Take for instance the example of one of the most widely advertised examples of a gated community: Palm Island. This artificially constructed island (in the shape of a palm tree) is located just off the shores of the city of Dubai and is nearing completion. It will provide a haven of luxury to those able to afford its villas and apartments.[4]

The ideology behind gated 'strawberry fields' communities such as Palm Island is remarkably similar to the one advocated by the European Union in its accommodation of wealth and its resistant, antagonistic and hostile practices when dealing with the mobile Other, especially deprived ones such as refugees, gypsies, immigrants, vagrants, and travellers (Urry 2000). Much like a gated community, the EU as a territorial system is designed to control, monitor and manage its borders and thereby protect those who are inside from those who are without. New members to the European club are sought after if they contribute to the sustenance or enhancement of economic prosperity, others are preferably stopped at the gates. Another group of people, unidentified yet of considerable size, slip through the maze, often with the help of human traffickers: they are the undocumented who clean and cater for the members of the gated community.

'Strawberry fields forever'?

Looking at the present geopolitical landscape of Europe, it can be ascertained that postmodern calls for and local celebrations of heterotopia notwithstanding, the making and marking of borders and thereby processes of social exclusion still survive in the EU. A wall is being erected that is increasingly fierce and formidable, yet also contains neoliberal mazes and is characterized by conscious blindness for specific (illegal) labour who help sustain ease and comfort. Neoconservative bordering practices not only increasingly fit the description of a gated community but also tend to reinforce capitalist and state-centred political logics. This occurs to the disadvantage of attempts to transcend neoliberal containment. Gated communities are a kind of never never land in the sense that the dream of purity and a life of ease is never-ending for, by definition, human desire is perpetual. Furthermore, inward-

---

4   A visit to the website of Palm Island is instructive: 'Welcome to a place like no other / A unique island experience / Ocean views and tranquil waters / Luxury properties starting at $275,000 / The Palm's horizons will forever change yours' <www.palmsales.ca>.

oriented 'strawberry fields' politics and outward-directed 'cherry-picking' policies appear increasingly untenable from a normative point of view. We would suggest that at a time when issues such as future EU enlargement remain controversial, or, as is the case with the Union's 'model of governance', hardly discussed at all (see, for instance, Kramsch et al. 2004), decision-making in the field of immigration and present European external border management practices are due for a radical overhaul. This does not, however, mean a further 'rationalization' of migration policies in terms of administrative capacities and logistics, but rather a return to a policy of solidarity and concern for the welfare of those who come from beyond the narrow confines of the European Union.

PART III
EU Enlargement and Its Impact at New External Borders

Chapter 5

Changing Border Situations within the Context of Hungarian Geopolitics

Zoltán Hajdú and Imre Nagy

A major focus of this book are recent geopolitical shifts that have not only reconfigured European borders but also transformed the nature of relationships between an enlarging EU and its immediate 'neighbourhood'. Hungary, an EU member since 2004, has undergone several dramatic transitions with regard to its geopolitical situation within Europe. Wedged between 'core Europe' and the Schengen area, on the one hand, and the emerging New Neighbourhood, on the other, Hungary's borders, and the management of these borders, will tell us much about the EU as an 'inclusive' and 'exclusive' political community. Several essays in this book deal directly with these questions, for example Béla Baranyi's which provides an overview of the evolution of cross-border cooperation at Hungary's eastern borders. In order to more clearly frame the overall geopolitical context of Hungary's 'EU-Europeanization', it is necessary to review historical developments that have affected central and eastern Europe during the past century.

Hungary's geopolitical status was fundamentally recast in the course of the twentieth century, with its territory and frontiers changing dramatically twice through war and its aftermath. This had a profound impact on its relations with its neighbours and its embeddedness in regional as well as continental power relations. After the second world war, Hungary – now as a small state – gradually adjusted to the geopolitical structures of a divided world and, within that, a divided Europe. Hungary's self-perception in geopolitical terms has also been dependent both on values promulgated by political elites and more general perceptions of 'neighbours', 'security', 'interstate community' and 'alliances'. The determinants of how a geopolitical situation is perceived domestically vary greatly from one social group to the next. The governing elite may, for instance, have a specific perception of the evolution of a country's foreign policy and its network of alliances and thus of the advantages or disadvantages of the country's geographical location. In terms of external perceptions of Hungary's geopolitical role this has entailed the evaluation of its participation in processes involving its neighbours and spatial community and in European and global trends. As Hungarian history attests, neighbouring states and European actors have several times shifted geopolitical positions and opinions on rather short notice. Hungary's geopolitical status and its changes can be examined and approached from several spatial aspects (Ring 1986; Pándi 1995;

Hajdú 1995; Romsics 1996; Pap and Tóth 1997), including 1) relations with its neighbours, 2) perception of its spatial community (for example, central Europe), 3) its functional large space (in-between Europe, a grey zone), 4) trends in Europe and 5) more global trends.

Although in terms of ethnicity, Hungary's neighbouring environment was relatively stable in the twentieth century, the importance and role of individual ethnic groups underwent dramatic changes. Voluntary and forced international migration as well as the spontaneous, often strikingly different, grassroots movements of various ethnic groups played a 'supporting' role in the transformation of the ethnic space (Kocsis and Kocsis-Hodosi 1998). In contrast, neighbouring states changed continuously, and on several occasions fundamentally, in the course of the twentieth century. New states (Czechoslovakia, Yugoslavia and the Soviet Union) have emerged and disappeared, neighbours have changed (Germany and Poland) and new states have become new neighbours (Slovakia, the Ukraine, Serbia-Montenegro, Croatia and Slovenia). Almost 'hidden' in the former Soviet Union, the Ukraine, for instance, hardly exerted any impact on trends in Hungary before December 1991. Since then it has become Hungary's largest, most populous and militarily most powerful neighbour. In addition, Austria's history in the twentieth century also had a major impact on Hungary's geopolitical situation (Lendval 1997; Kissinger 1994; Király and Romsics 1998).

This essay will provide an overview of Hungary's changing geopolitical status within Europe since 1919. These changes mirror more global events that have taken place at the European level. This essay will also provide background for several case studies presenter later in this book by the Hungarian authors.

Hungary and Geopolitical Determinants of the Twentieth Century

Global, continental and domestic developments as well as developments in neighbouring countries all influence the prevailing geopolitical situation of a given country and along its borders. In the twentieth century a constantly changing Hungary with constantly changing neighbours fitted in with the regional, economic and power structures in Europe. Within frequently changing European structures, Hungary and its neighbours belonged to one of the most unstable regions. With its boundaries redrawn radically and frequently, Hungary's geographical location has been defined on a number of occasions in the twentieth century and labelled as 'Eastern', 'Western', 'Southeast European', 'Balkan', 'Southern European', 'Southwest European', 'Central European', 'Central East European' and 'East Central European' (Ankler 1997; Beluszky 1995), each geographical characterization based on direct internal and external political considerations. Even Hungarians themselves were unable to arrive at a consensus on the country's geographical identity in Europe, perhaps reflecting the multifarious factors at work on Hungary's national identity.

Hungary's geopolitical characteristics changed from one era to the next in the course of its recent history, with various political ambitions proclaimed to its

Map 5.1   Hungary and its neighbours

neighbours. Before 1918, the Hungarian political elite and the rest of the society were sharply divided over the status of the Austro-Hungarian monarchy (a major power *vs* a minor power *vs* a power between the two), the security of its borders and the relationship between Austria and Hungary and the individual neighbouring countries. Prior to 1918, Hungary did not have any material border dispute with neighbouring states. However, some of the latter (that is, Romania and Serbia) laid overt claims to sizeable areas in Hungary (as far as the River Tisza and Lake Balaton, respectively). The Czechs living within the boundaries of the monarchy also envisaged a nation state incorporating large areas in Hungary (stretching as far as Budapest). Under the 1920 Peace Treaty, Hungary became an independent country on one-third of its former territory, with one-third of its population incorporated into neighbouring countries. Between the two wars it publicly voiced its claims for its former territories to its neighbours. There was a contradiction between its actual status (that is, that of a minor military power, a small country with a small population and low economic performance) and the feasibility of its ambition relying on its own resources. That is indeed why alliances became more important than before.

The perception of Hungary by its neighbouring countries was also rather mixed. The one that the population, the political elite and, in part, the scientific community in the successor states had after 1920 was based on the size, importance and opportunities of the pre–1918 Hungary rather than on its contemporary status. Essentially, successor states lost sight of actual power relations in shaping their foreign policy and were forging their alliances against Hungary. Between the two wars there emerged fundamental differences between the individual groups of pro-Italians, pro-Germans and the pro-British with regard to the perception of the country's status and the feasibility of its ambitions to annex its former territories. Though Hungary became Germany's neighbour after the Anschluss, it officially continued to lay territorial claims to Burgenland.

Before and during the second world war, parts of Hungary's former territories were annexed (the Hungarian Highlands in 1938, Ruthenia in 1939, Northern Transylvania in 1940 and the Voivodina in 1941), leading to a situation which proved transient and impaired the chances of subsequent cooperation with most of its neighbouring countries. The lack of a broad-based consensus manifested itself in rather a particular manner between 1945 and 1948, when the country, with its room for manoeuvre limited by the Russian occupation, took stock of its geopolitical opportunities. After 1949 Hungary adjusted to the Soviet sphere of interest. 'Fraternal, eternal and unbreakable friendship between Hungary and the Soviet Union' was (except for 1956) an unchallengeable tenet of official politics. Though this alliance and commitment were unacceptable for the majority of the Hungarian people, dissident ideas were repressed for decades. After 1949, furthermore, Hungary's geopolitical status was fundamentally determined by its inclusion in the Soviet alliance system in eastern Europe. The country became an integral part of the 'socialist camp', that is the Soviet sphere of political and economic interests. Surprisingly enough, it was economic relations that were first enshrined in an international treaty and institutionalized when, on 20 January 1949, the Council for Mutual Economic Assistance (COMECON), a council to promote economic cooperation between socialist countries (Bulgaria, Czechoslovakia, Poland, Hungary, Romania and the Soviet Union), was established.

Though the Warsaw Pact (1955) was supposed to be a military alliance, the military might of the minor signatories to the protocol was nowhere near that of the Soviet Union. The true underlying reason for the foundation of the pact was to enable the Soviet Union to have military control over the territories of its allies. Hungary bordered socialist countries along most of its borders (2,246 km or 84.1 per cent). From 1955, 56.9 per cent stretched along the 'borders of military alliance'. Before the Austrian state treaty was concluded in 1955, a mere 15.9 per cent of its state frontier ran along the borders of Austria, first under Soviet occupation and then a neutral nation. Under the contemporary system of values, political contiguity of this kind was perceived as favourable. (The perception of Yugoslavia or the shared borderline was far from being consistent or unambiguous at times.) In the era of bipolarity, bilateral relations between neighbouring countries were characterized by the coexistence of stability and instability. The very content of interstate relations played a key role in shaping border and cross-border relations.

In the 1950s, borders were in effect 'closed' along their entire length in terms of population relations. Differences only lay in the type and nature of the closure. After the 1956 revolution the geopolitical status quo along Hungary's borders became even more convoluted. Borders between allied countries were difficult to cross for most of the citizens. Though the Hungarian-Russian border was the scene of heavy freight and goods transportation, and military and ideological ties between the two countries were close, a sturdy barbed wire fence on the Soviet side of the border – the iron curtain – sent the unambiguous message that the Soviet Union was less than enthusiastic about encouraging the establishment of population relations on a massive scale. (The underlying reason for this was the ethnic Hungarians living on the Soviet side of the border.) After the withdrawal of the Russian troops from Austria (1955) a marked change occurred in the perception of the Austrian border. Border control was tightened on the Hungarian side and relations between the two countries were reduced to a bare minimum. After 1948 the border to Yugoslavia was practically closed in a military sense and remained so during the rest of the decade. Though the remaining borders were formally considered to be 'friendly borders', there was no political will to shape relations with the relevant countries.

The 1960s saw some material changes along the borders. Opportunities to establish better cross-border ties improved. Decisions were made on the level of official politics 'to develop cross-border ties with neighbouring friendly countries', and so there were plans to implement such decisions. The so-called 'red passport', introduced in Hungary and valid for travels in socialist countries, except Yugoslavia and the Soviet Union, allowed the majority of the Hungarian population freer movement and provided for the possibility of regular encounters and contact between Hungarian nationals and ethnic Hungarians living in neighbouring countries. From the second half of the 1960s increasingly special importance was attached to Hungarian-Austrian relations. Both parties were ready and able to demonstrate the advantages of the policy of peaceful coexistence of different economic and social regimes. The improvement in relations was due to the fact that Hungary had become the most presentable socialist country in terms of its internal affairs.

The Soviet invasion of Czechoslovakia in August 1968 (with minor allies providing a helping hand) was still 'within limits' tolerable to a divided bipolar world, and no countermeasures were taken by the West in response. Hungary's participation in the invasion unfavourably affected Hungarian-Czechoslovak cross-border relations. The 1970s saw improvements in the geopolitical sphere in nearly every direction. Détente and collective security efforts suited Hungarian interests perfectly. The ratification of the Final Act of the Conference on Security and Cooperation in Europe in 1975, in Helsinki, created favourable conditions for expanding manoeuvring room for Hungary's foreign policy.

Despite this, relations between Hungary and its neighbours were far from good in the 1980s. In fact, those with Romania became rather strained, mainly due to the rapid deterioration in the treatment of ethnic Hungarians (Shafir 1985). Hungarian-Czechoslovak relations also had their ups and downs, primarily because of the planned construction of a dam on the River Danube at Gabcikovo-Nagymaros and the treatment of ethnic Hungarians (Wolchik 1991). But Hungarian-Yugoslav relations were, compared to earlier decades, relatively smooth. Yugoslavia was busy sorting out its own problems, facing the challenges and contradictions of change in the post-Tito era. In the final decade of state socialism the Hungarian economy became an integral part of COMECON, and the Soviet division of labour in particular. Socialist economic 'commitment' was gradually abandoned through the advocacy of national interests and integration into the Western economy.

Based on total turnover the fundamental structure of Hungarian foreign trade in 1985 reflected the dominance of trade with socialist countries, and within this COMECON countries in terms of both imports (54.4 per cent) and exports (58.6 per cent); the proportion of developed capitalist countries was significant (38.5 and 30.8 per cent, respectively) while that of developing countries was low (7.1 and 10.6 per cent, respectively). For economic and political-ideological reasons Hungary's most important trading partner was the Soviet Union, having a share of 30 per cent of imports and 33.6 per cent of exports. Hungary's second-most important trading partner was the Federal Republic of Germany, with a share of 11.4 per cent in exports and 7.8 per cent in imports.

The development of cross-border ties involving areas along the western and southwestern borders in the second half of the 1980s was a new phenomenon. First as observers, then as members, the counties lying along Hungary's western border joined the Alpine-Adriatic Working Group and became increasingly involved in its activities. The significance of this working relationship was far-reaching. Attempts at comprehensively developing border and cross-border with socialist allies failed. Already by 1989 the 'non-rouble based' turnover was dominant (61.6 per cent), while the proportion of turnover settled in roubles shrank (38.4 per cent). A similar 'shift in the pattern of financial settlement' was discernible in exports, with 62.2 per cent of all exports being settled in other currencies. In 1989 the proportion of foreign trade (imports and exports) was as follows: COMECON countries, 41.8 per cent; countries of the European Economic Community (EEC), 29 and 24.8 per cent; countries of the European Free Trade Area (EFTA) 13.8 and 10.7 per cent; and the rest of the world, 17.6 and 27 per cent respectively. Largely due to the Soviet Union, Hungarian foreign trade turnover was as a rule centred on neighbouring countries, hence on Europe.

As regards military structures Hungary was fully incorporated into the Warsaw Pact. In effect, it accepted the military doctrine and supremacy of the Soviet Union. With its operation subordinated to the Warsaw Pact, the Hungarian army was wholly dependent on the Soviet Union for such things as the supply of military technology and the planning of military operations. In essence, military policy in Hungary was subject to the status that it had in the territorial system of the Warsaw Pact with its perceived 'enemy image' and in Soviet geostrategy. Within the system of military alliance Hungary was classified as a Southwest European country due to its strategic location. It was not in the front line of the division between the two regimes. Rather it was defined as a country lying in 'the direction of minor military operations'. (Yet the amount of military technology amassed, discovered upon the withdrawal of the Soviet troops from Hungary, pointed to meticulous preparation for major military offensives in this direction as well.)

The stationing of the Soviet Southern Group of Forces in Hungary served the purposes of not only external but also, if needed, internal oppression. The territorial layout of the Soviet military bases unequivocally pointed to the set-up of a battle order in a westerly direction; the Russian ring around Budapest was a key component of army force allocation (possible internal oppression). The Hungarian army was larger than the Soviet forces stationed in Hungary in terms of headcount. However, it was significantly weaker in terms of fire power. The Hungarian air force was especially weak and outdated compared to its Soviet counterpart. The military structure meant that the country and the Hungarian army were both vulnerable. Hungarian-Soviet military relations remained stable until the political changeover. In January 1988 a military exercise called 'Friendship 88' was mounted with the participation of Hungarian, Czechoslovak and Soviet forces. In line with new European agreements and treaties governing issues of politics, security and military policy, observers from the West followed the exercise.

In October 1988 Budapest hosted the meeting of the Military Council of the Armed Forces of Warsaw Pact member states, which examined the possible outcome of a reduction in armed forces. The meeting was followed by a military exercise with the participation of Hungarian and Soviet forces. Observers from the West also attended the exercise. In December 1988 Mikhail Gorbachev announced at the general assembly of the United Nations that the numbers of Soviet troops in Hungary, Poland and Czechoslovakia would be reduced. Accordingly, the defence ministers of Warsaw Pact countries afterwards only talked about 'the provision of adequate protection' at their meeting in Budapest in November 1989. This allowed for the possibility that the Warsaw Pact could actually be transformed into a defence organization. After 1988 an interaction between a protracted internal crisis in the Soviet Union with an increasingly weakening status in global politics (and also a diminished position as a superpower) and the internal transformation of neighbouring countries brought about domestic developments. This interaction opened up new possibilities for Hungary's geopolitical status to change (Agnew and Corbridge 1989).

Of all socialist countries, except for the German Democratic Republic which had a special status due to trade with the Federal Republic of Germany, Hungary was the first to realize the importance of the European Economic Community. As early as 1982 Hungary started expert-level negotiations with the EEC to improve trading conditions. The expanding EEC played an increasingly important role in Hungarian foreign trade. Hungary had a vested interest in developing bilateral economic relations in all spheres. Examined from the perspective of 'statistical ease of crossing borders', neighbourly ties reflected a particularly unique situation 'during the final years of building socialism'. Based on the average length of border per road or railway border crossing station at the end of the era and from nearly all other perspectives after 1988, the Austrian border was the easiest and the Soviet one the hardest to cross. With the introduction of the so-called 'world passport' in 1988, swarms of Hungarians went on shopping sprees to Austria. Still, Hungary was not allowed to ignore the security expectations of the 'socialist camp'. As a result it was not until the end of the era that, for instance, the 'iron curtain' was dismantled.

The two countries agreed on putting an end to the institution of sealed borders in February 1989. It only took a few weeks to dismantle them physically.

From 1986, completely lawful under the constitution, various dissident groups and organizations were established, with rethinking the country's geopolitical status as one of their major objectives. Though some still accepted the political shelter offered by the Patriotic Popular Front, manifestos concerning the country's international status contained radically new demands:

- return to the European fold (the reintegration of Hungary as an independent nation into the political, economic and cultural community of European nations);
- integration of central Europe (nations in the region should enter into economic and political alliances based on mutual advantage of their own accord and without any external intervention; this would enable ethnic issues to be solved in a democratic spirit);
- a unified Hungarian nation (all 15 million Hungarians should belong to a unified Hungarian nation; the government in office should support the cause of ethnic Hungarians on various international fora);
- an independent Hungary (based on contemporary national interests, Hungary should re-evaluate the system of international relations established during the Stalin era).

Operating as parties (the Hungarian Democratic Front, the Alliance of Young Democrats and the Alliance of Free Democrats), though not yet parties in name, social movements increasingly gained strength. These political movements were able to tackle the issue of the direction of the country's foreign policy without having to observe any past commitments. Party manifestos and official statements in 1989 centred primarily on reviewing the alliance system to which Hungary belonged as well as on the consideration of opportunities and limitations. Each major political movement faced the issue of the Warsaw Pact and COMECON. Under the realpolitik approach most parties set the overhaul of these organizations as a minimum goal, and Hungary's 'approved' exit from them as the ultimate objective. Most political parties were supportive of Hungary becoming a neutral state. Many articulated the need for the simultaneous dissolution of NATO and the Warsaw Pact, and the establishment of a collective security system in Europe. In June 1989 the government in power and the various dissident groups started negotiations, called the National Round Table Talks, over the country's current state of affairs and future. At stake was a peaceful transformation that represented a happy medium. Though negotiations focused mainly on home affairs, the issue of the direction of the country's foreign policy was also raised.

By allowing East German asylum seekers to leave the country, the Hungarian government unequivocally abandoned an earlier practice adopted in socialist countries. Pulling down the iron curtain had international reverberations and sent shock waves through East Germany. At its Fourteenth Congress held on

6–10 October 1989, the Hungarian Socialist Workers' Party (MSZMP) dissolved itself and established the Hungarian Socialist Party (MSZP) as its successor. The dissolution of the former state party created a new situation in which the MSZP was forced to enter the political fray, encumbered by the chequered political past of its leaders. Democratic changes in the socialist countries in Europe called into question former economic, political and military structures as well as the unity of the socialist camp.

A new feature of foreign policy manoeuvring was the possibility of deepening international cooperation between Yugoslavia, Italy, Austria and Hungary. In November 1989 Austria, a neutral country, Yugoslavia, a non-allied one, Italy, a NATO member, and Hungary, a Warsaw Pact state, held high-level discussions on the possibility of closer economic and political cooperation between the countries in the region. Simultaneously with this initiative the Hungarian government submitted its membership application to the Council of Europe. In March 1990 the Hungarian and Soviet government agreed on the full withdrawal of the Soviet troops stationed in Hungary, which began almost immediately after the signing of the agreement. This sent a message to the public about home affairs before multi-party elections, namely, that the new MSZP was able to restore national independence and represent the country abroad.

Foreign policy and Hungary's potential neutral status were key issues in the 1990 election campaign. A survey of the views that the various political parties adopted on foreign policy and alliance policy reveals that they fully agreed that Hungary's independence be restored and that it had to return to the European fold. In April 1990, marking the birth of what was to become the Visegrád Group, the prime ministers and heads of state of Czechoslovakia, Poland and Hungary held discussions on increasing cooperation between fledgling democracies. The fundamental issue was for westward-lying, more developed members of COMECON and the Warsaw Pact, the flagships of democratic transformation and with Poland as their potential leader, to concert their actions (Story 1993).

Hungary's Geopolitical Status in the 'Grey Zone' between 'East' and 'West'

In parallel with economic, social, political and power restructuring at the national level, there was a transformation of the former global status quo as well: European structures changed radically, and the socialist alliance system collapsed. The 'grey zone' in their wake meant new room for manoeuvre, some sort of freedom, responsibilities, challenges and concerns for Hungary – all at the same time (Tunander et al. 1997). The Hungarian Democratic Forum won the 1990 multi-party elections, though not with an absolute majority, and formed a coalition government. Soon after the coalition was established and before the new government took office, the new Parliament put the issue of reviewing and deciding on the future of the relationship between Hungary and the Warsaw Pact on its agenda. Parliament dismissed a unilateral exit and requested the future government to start negotiations on Hungary's leaving the alliance.

As regards foreign policy, the government programme adopted the approach of realpolitik inasmuch as it declared that it would abide by earlier international treaties; at the same time, consistent with the new political status quo, it set new goals, prioritizing the issue of European integration. After the elections, cooperation in the Adriatic-Danube region (Pentagonale) was re-emphasized. This 'in-between blocs' formation played an important role in maintaining stability in the region and won direct political support from the West for a democratic changeover (Bierman and Loboda 1992; Carter 1995). Hungary joined EFTA (June 1990), established diplomatic relations with NATO (July 1990), and the European Economic Community opened an embassy in Budapest (November 1990). The establishment of diplomatic relations between NATO, the EEC and Hungary, still a Warsaw Pact and COMECON member at the time, was tantamount to the recognition of and support for the new government's foreign policy.

In September 1990 the Hungarian government approved East Germany's desire to leave the Warsaw Pact. Given the German unification, all things considered, this move was a logical one and reinforced Hungary's legal stance that it was possible to withdraw from the treaty. In October the Hungarian government confirmed its intention to leave the military organization of the Warsaw Pact (which also meant that the issue of full withdrawal from the organization had not yet been raised). In January 1991, through deploying a health contingent, Hungary participated in operation 'Desert Storm', aimed at ousting Iraqi invaders from Kuwait. Though the operation was conducted under the aegis of the UN, it drew on US military might. Through its participation in the operation, Hungary was able to demonstrate its newly gained international leeway as well as its commitment to western values (Balogh 1998; Bognár 1993).

The new government focused on the issue of foreign national ethnic Hungarians as never before. The prime minister József Antall declared that he wanted to be 'prime minister of 15 million Hungarians in spirit', which marked the beginning of a new national policy, one that foreign policy was to implement. Neighbourhood policy was nearly synonymous with addressing the issue of ethnic Hungarians, particularly those living in Romania. Diplomatic relations having been established, ties with the EEC and NATO became increasingly close. It was not an insincere policy of currying favour with both organisations but rather one of exploring new opportunities in an intricate international and domestic situation (Sisposné Kecskeméti and Nagy 1995). Neither the United States nor Western Europe had been prepared for changes of this kind and the rapidity with which they occurred in the East.

The issue of the dam on the River Danube at Gabcikovo-Nagymaros was still haunting Hungarian-Czechoslovak relations, which further deteriorated when in January 1991 Czechoslovakia announced that it would go on with the construction. This resulted in diverting the river course of the Danube. Though the dispute was submitted to the International Court at The Hague, the court ruling alone failed to ease the tension between the two countries representing two contrasting approaches and objectives. In February 1991 Czechoslovakia, Poland and Hungary signed an agreement of cooperation. Cooperation between the Visegrád countries attempted to fill a political and power vacuum. However, Poland was too weak to become an internationally recognized leader in the region.

At its extraordinary meeting in Budapest on 25 February, the Political Consultative Committee of the Warsaw Pact abrogated earlier military treaties and disbanded the military organization of the pact with effect from 31 March 1991. On 19 June the last Soviet soldier left Hungarian soil. On 28 June 1991 Hungary approved the dissolution of COMECON, thereby creating new economic conditions under which to operate not only in respect of former member states but also other regions. A few days later a resolution was passed on the complete dissolution of the entire Warsaw Pact – not just its military organisation. On 1 July 1991 Hungary became an independent nation, outside any bloc, any integration and with no foreign military presence on its soil.

Rising tensions in the Balkans, compounded by signs of a civil war from August 1991, made Hungarian–Yugoslav relations even more strained. Debates were especially bitter over Hungary's arms transports to Croatia and, later, over Yugoslav skirmishes along the border (Owen 1995; O'Loughlin and van der Wusten 1993). Simultaneously with the transformation of its geographical surroundings, Hungary and the EEC ratified a treaty in November 1991, under which Hungary became an associate member. Though this did not automatically guarantee Hungary's actual membership, it could be construed as the first step towards it. An aborted coup d'état in the Soviet Union in August 1991 and the subsequent official dissolution of the Soviet Union in December that same year created a fundamentally new situation for Hungary. With the Ukraine becoming independent and a new Russia coming into existence, new state structures, power relations and spheres of interests emerged along Hungary's northeastern border and in its wider eastern region (Nagy 1995).

New priorities in neighbourhood policy were clearly reflected in Hungary's being among the first countries to acknowledge the independence of Croatia and Slovenia on 15 January 1992 and, later, to establish diplomatic relations with these two new neighbours. Though Hungary repeatedly emphasized its neutral stance in the Balkan crisis, largely due to ethnic Hungarians in the Voivodina, it effectively took sides with Croatia and Slovenia, supporting their efforts whenever possible. It was not only this siding, but also Hungary's support for the punitive measures initiated by the international community against Serbia that contributed to an increasingly troubled Hungarian-Serbian relationship. In the course of overseeing compliance with punitive measures, new relations were established between Hungary and West European institutions of integration. (Though compliance with punitive measures caused Hungary economic loss, political gains outdid it. Economically speaking, the true winners were Hungarian and Yugoslav smugglers living near the border.)

The Hungarian government sought to conclude framework treaties with its neighbours, Soviet successor states and major partner countries, to consolidate relations with them. The first treaty of the kind was the Hungarian-German framework treaty, signed in February 1992. Framework treaties followed between

Hungary and Lithuania and Hungary and Croatia. The one between Hungary and the Ukraine was highly contentious, leading to a crisis within the governing coalition, so much so that the government had to win over the opposition to get the treaty passed. The dissolution of Czechoslovakia in 1993 resulted in the emergence of a new state and a new situation along Hungary's northern border as well. The proportion of ethnic Hungarians in the new Slovak state grew, which, rather than easing the tension, made relations between the two countries even more discordant. All in all, the landscape of the bordering countries was fundamentally reconfigured in the early 1990s, with Hungary's position unambiguously strengthened relative to those of the new successor states. The Hungarian society and the overwhelming majority of the political elite exercised self-restraint in connection with the transformation processes in neighbouring countries. So did the majority of ethnic Hungarians living there.

The Central European Free Trade Association (CEFTA) with some of the neighbouring countries created new opportunities and forms of cooperation. At the same time, however, especially in respect of agricultural exports, it was also the scene of open clashes of interests, which were rather clumsily handled. Nevertheless, CEFTA has played a positive role in the system of relations between former socialist countries, as it has been instrumental in internalizing and addressing some of the likely outcomes of EU integration.

Hungary as NATO member

Enjoying a majority in Parliament, the social-liberal coalition government, which came to power after the 1994 elections, stood for the continuation of earlier foreign policy. Perhaps the only exception were relationships with neighbouring countries, where a change in attitude materialized, if not a new trend. The new framework treaties between Hungary and Romania and Hungary and Slovakia (though not welcomed by all domestic political parties) were aimed at consolidating bilateral relations and promoting Hungary's ambitions to join NATO and accede to the European Union. The new coalition government resumed and even accelerated preparations for both NATO and EU membership (Bognár 1995; Burdack, Grimm and Paul 1998; Deák 1997 and 2004).

Given the new global and continental status quo, imminent NATO enlargement raised the issue of the security of the Ukraine and Russia. Launched in January 1994, the Partnership for Peace programme reflected a need for greater trust. If NATO were to prove that it did not perceive the two militarily strongest Soviet successor states as enemies, it had to involve them in a common security system. The special relationship with the two countries governed by the treaties granted the two Soviet successor states a minimum guarantee (Ehrhart 1997). Smaller ex-socialist countries competed with each other for inclusion in the first round of NATO enlargement. For the entrants, membership symbolized an external guarantee of security and was the high-level recognition of their internal democratisation. (Janning and Weidenfeld 1993)

The NATO decision in July 1997 to invite the Czech Republic, Poland and Hungary to join the organization was the manifestation of an internal compromise reached by the then member states. Nevertheless, it was also the recognition of democratic development in the three countries. Following a successful referendum and with the ratification procedures completed, Hungary became a NATO member in the spring of 1999. Given the contemporary political and military status quo and an intricate sphere of interests, Hungary's NATO membership preceded its accession to the European Union. (At the very beginning of the internal transformation process, an opposite scenario had seemed more likely.) Hungary's status was rather special within NATO's territorial structure. It became a 'land-locked island', which meant that it had no common border with any other member state. This 'isolated' status did not create any problems at the time, since Slovenia continued to follow a NATO-oriented policy, and after the 1998 general elections, Slovakia also stepped up its preparations for NATO membership. (Domestic debates in Austria over eternal neutrality *vs* NATO membership will probably continue for long years to come.)

NATO's air strike against Serbia in the spring of 1999 put Hungary – a new member of NATO – in a rather awkward situation. It indirectly became party to a military intervention (in effect, an undeclared war waged without UN authorization) against a neighbouring country. However, in terms of its NATO membership, it was very direct participation. Air strikes were launched from its territory and through its air space against a neighbouring country.

Hungary is under NATO's southern command, which adds to Slovenia's importance, and is likely to have further ramifications. NATO member Hungary lies in the vicinity of the Bosnian trouble spot, with an air base in Taszár (also a major logistics centre) playing a key role in providing supplies for international forces in Bosnia. NATO membership does not formally conflict with the country's sovereignty. (Not all military and political ramifications of alliance commitment follow from the fact of being part of an alliance.) Yet in practice, under a new alliance system, relations are placed in a new perspective and issues of security are raised differently. In the course of the NATO enlargement in the spring of 2004 Slovenia, Slovakia and Romania also joined the organization. As a result, Hungary ceased to be a 'land-locked island', with the majority of its borders once more becoming 'inner borders shared with allies'. It also added to the country's security, with the majority of ethnic Hungarians in neighbouring countries included in the same system of alliance and democratic values.

Hungary as a Member of the European Union

The late 1990s marked the arrival of a new era in the history of the European Union. It had to address the issues of deepening integration, the introduction of the single currency and take stock of the experiences of its operation. This had to be done in circumstances where new desires to advocate national interests within the EU surfaced (Hargital, Izikné Hedri and Palánkai 1995; Inotai 1998; Izikné Hedri

1995). There emerged several scenarios of EU enlargement (in small, medium-sized or large groups) (Weidenfeld and Altmann 1995; Urwin 1999). Initially, it was the small group scenario (the Czech Republic, Poland and Hungary) that looked likely. At its summit in Luxembourg in December 1997, the EU identified the first round of countries to be invited to participate in the accession talks. Hungary was involved in the five-plus-one configuration. (At the time the medium-sized group scenario looked viable.)

Not only the identity of the potential entrants but also the dates of their accession were unknown for a while. There had been several accession dates suggested during the talks before the May 2004 date was finally fixed. In practice, the large group scenario prevailed, allowing ten countries to simultaneously accede to the EU. After accession, Hungary's borders took on a dual aspect. The majority of its borders became inner borders of the EU (Hungarian-Slovenian, Hungarian-Austrian and Hungarian-Slovak); at the same time the Hungarian-Ukrainian, Hungarian-Romanian, Hungarian-Serbian and Hungarian-Croatian stretches became the EU's outer borders with all the associated legal, economic and political consequences. Romania has been promised entry in 2007, and Croatia's rapid accession cannot be ruled out either. Consequently, for Hungary outer EU borders may lose their importance.

Because the issue of Hungary's EU membership was put to a referendum, it was the voters' decision rather than the exclusive ambitions of the political elite that legitimized accession. Essentially there was general consensus on accession among voters interested in political developments. Yet the low turnout at the referendum should definitely be perceived as a warning sign for the future. As the overwhelming majority of Hungary's foreign trade relations was already EU-oriented prior to its accession, accession itself, despite seen and unseen risks facing both parties, was not a turning point in this respect. On the contrary, it consolidated the structures that had already been put in place. Even though the process of constitutional legislation in the EU had ground to a halt by the spring of 2004, compromise was reached on weighted votes in the summer. (In several countries, disputes over the need for referenda on the European constitution added further uncertainty to the process.) It is also in Hungary's long-term interests that it should participate in the integration processes of a functioning EU with a stable and transparent structure and clear objectives.

A NATO member and an EU member state, Hungary has become an undisputed part of the system of Euro-Atlantic integration and processes, thereby transforming itself from 'the West of the East' into 'the East of the West', and not just topographically. However, the historical geopolitical determinations and characteristics of the neighbouring countries will have to be reckoned with in the future. Austria is an EU member, but a neutral state. Croatia is on its way to EU and NATO membership, though the date of its accession has not yet been set. And while Serbia-Montenegro has formally declared its intention to join the Euro-Atlantic system, its accession may be a long drawn-out process. As for the Ukraine (sandwiched between the EU region and Russia), the search for political stability and good relations both to Russian and the EU could require considerable time.

Cross-Border Cooperation: Some Remarks

This overview of Hungary's changing geopolitical situation indicates that the country is a 'borderlands nation'. There is no region within Hungary that is not affected by events in neighbouring states or by cross-border flows of goods, capital or persons. With Hungary's almost total integration into 'Western' institutions and the end of regional military strife in the Balkans, Hungary's borders have not only become open but also conducive to regional cooperation. With increasing opportunities and incentives for cross-border cooperation, Hungary's desire to 'spiritualize' the dividing post-Trianon borders is now close to fulfilment.

As several essays in this book indicate (see, for example, Béla Baranyi), the development of Euroregions at macro-, meso- and microregional levels can be seen as an attempt to capitalize on new potentials for political dialogue and to more properly address questions of minority rights in central and eastern Europe. The perspectives for regional economic and social development through cross-border cooperation should not be underestimated. Complementary settlement networks and development interests as well as a common desire for greater stability and prosperity within the EU framework provide powerful incentives. Euroregions, now located along all Hungarian borders, are a further indicator of integration into EU governance structures. Initial fears that Euroregions might be used as political tools for realigning present state borders have all but dissipated. The geopolitical view has shifted from bilateral relations to a more integrating European one. The partial 'Europeanization' of national geopolitical discourses and practices will, of course, not resolve all contentious issues Hungary and its neighbours must deal with as they intensify economic, political and social interaction. However, it can be assumed that the borders will no longer play as divisive a role as in the past.

Conclusions

Hungarian foreign policy has always had to be realistic owing to the country's geographical location and the processes of global, continental and neighbourhood restructuring. Based on Hungary's national as well as domestic and foreign policy failures and tragedies in the twentieth century, one may easily arrive at the conclusion that Hungary's foreign policy has always 'erred', judging long-term developments in external conditions erroneously. However, such a judgement would only be partially true. The majority of the neighbouring countries underwent similar tragedies the past century. Hence, the true facts have more to do with regional and continental characteristics than country-specific problems (such as the incompetence or impotence of politicians or those involved in foreign policy).

Hungary's opportunities changed radically in the new status quo that had evolved by the late 1980s. In 1990 Hungary found itself in a completely new setting. With the Warsaw Pact disbanded, the Soviet troops gone, COMECON dissolved, along with internal social, economic and political changes, the dismantling of the former political regime and the political changeover, new opportunities opened up for Hungary on shaping its internal status quo and international relations. In the course of the decade, owing to internal and external changes, state socialism was first called into question, then failed at the 1990 democratic multi-party elections. This date marks the most important turning point in the era, since it created Hungary's independence and non-alignment to military blocs.

Not only Hungary's global environment, but also its relations with neighbouring states and wider Europe have markedly changed. It established relations with West European organizations of integration well before fundamental social and political changes occurred. In the new situation the Hungarian political leadership and the general public set accession to the changing European Union as an ultimate objective and value, and started deliberate preparations for accession and an envisaged membership. The changes in Hungary's geographical surroundings were characterized by the dissolution of former socialist federations and an increasing number of independent neighbouring countries. In many respects, newly independent neighbouring countries (Slovakia, the Ukraine, Croatia and Slovenia) carried less weight than their respective predecessor states did. Owing to this, Hungary's room for manoeuvring increased. The process of the emergence of new states in Hungary's neighbouring environment caused uncertainty. In particular, civil wars in the Balkans posed challenges to security. At the same time, it was these harrowing events that accelerated the process that finally led to Hungary joining NATO.

Joining the Euro-Atlantic military alliance greatly transformed the country's external security policy conditions. Although Hungary was now under the protective umbrella of European stability, the air strike against Serbia proved that the new alliance system also carried political risks and that there were embedded limits of advocacy of interests. With its EU membership the path of integration that Hungary had previously followed took on a new meaning. However, it should also be kept in mind that accession marks the beginning of a long process of accommodation, adjustment and learning, and that, in the short term, certain sectors, social layers and groups may well be the losers of new structural changes. As for the Hungarian nation that has lived through so much, the manner of joining NATO and accession to the European Union can be judged a historic turning point. In the course of its modern history, this time it was of its own accord and with its intentions legitimized by a referendum (rather than out of necessity) that the country joined a new alliance and became part of a new integration process.

Chapter 6

The Impact of EU Enlargement on the External and Internal Borders of the New Neighbours: The Case of Ukraine

Olga Mrinska

Even after enlargement in 2004 moved the EU's borders to the eastern edges of Poland, Slovakia, and the Baltic States, much of Europe has remained outside the Union. As Bulgaria and Romania are preparing for membership, several other nations of southeastern Europe still have a long way to go before joining the EU. Further east, Belarus, Ukraine, Moldova and much of Russia form a distinct part of Europe's architecture, with unique cultures that are in many ways quite different from those farther west. These differences have become stronger and have been sharpened during seven decades of socialism. Yet in other ways and at other times, these nations have had a common history with the rest of Europe, and in particular with some of the new member states. Overall, they are undeniably European.

For governments of these 'new neighbour' states as well as for their citizens, the future shape of Europe has become a crucial question: how will the expanded EU deal with them? Some fear that protective measures against perceived security threats to member states, such as those stemming from illegal immigration, organized crime, drug trafficking or weapons proliferation, might transform the EU's borders into a new iron curtain. Such an outcome appears, however, unlikely: building new barriers through closed borders and new regulations would widen the political and economic gulf between the EU and its eastern neighbours, and the isolation of 'outer Europe' might conceivably lead to increased political instability beyond the EU's borders and to hostility towards a 'European way of life'. Many more people therefore expect the EU to raise political dialogue with its neighbours to a new level, by offering them better terms during negotiations, by taking into account common values as well as specific circumstances and by building on already existing formal and informal links, especially between new member states and new neighbours. Indeed, this would be a continuation of the dynamic processes of integration and policy harmonization which EU policy has promoted during the past fifteen years, thereby acknowledging the vital importance of supranational decision-making and the need for reaching consensus to pursue the members' own interests. The accession of ten new states to the EU in 2004 has thus strengthened Europe's potential to act as an equal power to the US in the global economy and in world politics, but the realisation of this potential will require constant efforts over the coming two decades to build a secure and stable political and economic continental system from which the EU's new neighbours cannot be excluded.

The introduction of the European Neighbourhood Policy (EPN) has been a first step in this direction, demonstrating the EU's readiness to 'share the benefits of enlargement with neighbouring countries' by extending cooperation to a wide range of areas (the rule of law, good governance, security, respect of human rights, principles of a market economy and sustainable development). Although these areas clearly reflect priorities of security and economic stability meant to benefit member states as much as their neighbours, there is much truth in the EU's claim that ENP offers its 'ring of friends' the chance to share 'everything but the institutions'.

Ukrainian leaders have repeatedly stated their country's future to be European, and the recently elected government has just reaffirmed its long-term aspirations to join NATO and the EU. And whereas, in the past, Ukraine's European ambitions have been hampered by the perception that its attempts to build a free-market economy and a sound democratic state were not whole-hearted, the new leadership is firmly committed to implement major political, economic and social reforms which will help accelerate the integration process, such as the development of mechanisms that will allow to gradually bring in line Ukrainian legislation, institutions, and practices with prevailing EU standards in such areas as trade, finance, industry, services, transport and energy, infrastructure, education, and research and development.

However, it must be stressed that – just as has been the case with the countries of central Europe – the prospect of eventual EU membership will be vital for the stimulation of enduring and sustainable political, economic, social, military and judicial reforms. The current lack of a clear signal from the EU as to whether Ukraine might one day be accepted into the Union has thus been reinforcing euroscepticism in the country and pushing the government to favour initiatives promoted by the Commonwealth of Independent States and the Single Economic Space. In fact, attitudes towards future membership in NATO and the EU are well divided, with western and central Ukraine tending to be pro-European and eastern and southern Ukraine pro-Russian. According to a recent sociological survey, people in the eastern and southern parts hold almost identical views about the country's future political and economic development: they favour integration with Russia and other republics of the former Soviet Union and Russian as the official language, are hostile towards the EU and NATO and possible integration into them, do not believe that entrepreneurship can be an engine of economic growth, and appear to lack personal initiative relying instead on decisions made by the state. People in western and central Ukraine, on the other hand, tend to be strong advocates of private property (for both private and business purposes), see the country's future as being closely linked with the EU and other European institutions, are critical of Russia's attempts to regain influence in the region and reject the dominance of the Russian language (Democratic Initiatives Fund and Kyiv International Institute of Sociology 2005).

To some extent, this dichotomy reflects patterns of social and economic transformation as well as the present growing imbalance between different regions, due in part to the fact that Ukraine now shares a border with the EU. Historically, the country has been divided along a north-south axis, the so-called Huntington Line, with the River Dnipr forming a sort of natural border. Western Ukraine has shared long periods of history with Poland, Slovakia, Hungary and Romania, a common heritage which has resulted in social and cultural similarities as well as continuing informal relations across state borders that have only been established a few decades ago. Eastern and southern Ukraine, on the other hand, were under the strong influence of the Russian Empire for several centuries, and people in these regions still share numerous attitudes and cultural features with today's Russians.

Regional Disparities – the Economic and Social Context

Diverging general attitudes and foreign policy preferences can thus be traced back to different regional histories and the resulting different ethnic composition of Ukraine's regions. Recent fundamental economic and social changes have added still further to these disparities, notably since about 1999 when the national economy entered into a period of stable economic growth, with GDP growth rates of 9.4 per cent in 2003 and 12.1 per cent in 2004. Some regions have been able to adjust rather well to a competitive market economy, while many others continue to strongly rely on state subsidies even for sectors in which local or regional stakeholders could be expected to play a major role. For present purposes, Ukraine's 25 *oblasts* (provinces) can thus be divided into eight macroregions of homogeneous economic structure and common social characteristics.

The *Donetsky and Prydniprovsky regions* are long-standing industrial centres with a highly developed heavy industry sector and industrial output amounting to a quarter of total production. They currently account for over half of the nation's exports and for a third of added value in an economy where 45 per cent of all exports are produced by metallurgical and chemical industries (State Statistics Committee 2004). The economic potential of these sectors has, however, been gradually diminishing as they are using resource-intensive technologies damaging to the environment and of low added value. Their present competitive advantage in the international market might therefore be lost during the coming years, plunging currently well-off companies into a deep crisis and leading in turn to a severe economic decline in these regions. The recent ban on sectoral state subsidies and the end of tax exemption have added to these difficulties. It is also worth noting that these industrial centres receive less than a quarter of foreign direct investment, which is mainly directed to sectors allowing for quick profits, such as trade, services, the food and light industry, machinery and energy. Unless radical steps are being taken to bring these industries in line with international standards or to diversify industrial output, the economic prospects for these regions, and for the country as a whole, look bleak (Mrinska 2002).

The *eastern region* was one of the major producers of machinery during the Soviet era. But the share of machinery in exports, formerly a high political priority,

Table 6.1   Ukrainian regions and their relative economic weight, 2003

Macroregion oblast	Territory (in %)	Population (in %)	Gross value added (in %)	Industrial output (growth in 2002)	Foreign direct investment (in %)	Export of goods in 2004 (in %)
Ukraine	**100**	**100**	**100**	**116**	**100**	**100**
Donetsky	**8,8**	**15,2**	**16,5**	**–**	**7,4**	**31,3**
Donetsk	4,4	9,9	12,4	119	6,5	25,5
Lugansk	4,4	5,3	4,1	112	0,9	5,8
Carpathian	**9,3**	**13,0**	**9,2**	**–**	**9,0**	**6,9**
Zakarpatska	2,1	2,6	1,6	144	2,7	1,9
Ivano-Frankivsk	2,3	3,0	2,3	128	1,4	2,7
Lviv	3,6	5,5	4,3	114	4,6	2,0
Chernivtsi	1,3	1,9	1,0	127	0,3	0,3
Southern	**18,8**	**15,3**	**12,7**	**–**	**11,3**	**6,7**
ARC	4,5	5,1	3,7	119	3,8	0,8
Mykolayv	4,1	2,6	2,2	112	1,1	2,0
Odesa	5,5	5,1	5,2	118	5,4	3,2
Kherson	4,7	2,5	1,6	116	1,0	0,7
Podilsky	**10,1**	**8,8**	**5,7**	**–**	**2,1**	**2,2**
Vinnytsa	4,4	3,6	2,5	112	1,0	1,3
Ternopil	2,3	2,3	1,3	135	0,4	0,3
Khmelnitsky	3,4	2,9	1,9	114	0,7	0,6
Polisky	**16,9**	**10,0**	**6,7**	**–**	**4,4**	**3,0**
Volynska	3,3	2,1	1,4	122	1,3	0,8
Zhytomyr	5,0	2,9	1,7	122	1,2	0,8
Rivne	3,3	2,4	1,7	121	0,8	0,6
Chernigiv	5,3	2,6	1,9	110	1,1	0,8
Prydniprovsky	**13,9**	**13,8**	**14,5**	**–**	**16,5**	**24,5**
Dnipropetrovsk	5,3	7,5	8,7	110	9,0	16,6
Zaporizhzhia	4,5	4,0	4,2	121	6,6	7,3
Kirovograd	4,1	2,3	1,6	124	0,9	0,6
Eastern	**13,9**	**12,1**	**11,5**	**–**	**8,7**	**7,9**
Poltava	4,8	3,4	3,6	109	2,6	4,2
Sumy	3,9	2,7	2,0	117	2,1	1,4
Kharkiv	5,2	6,0	5,9	111	4,0	2,3
Central	**8,3**	**12,0**	**23,2**	**–**	**40,6**	**15,4**
Kyiv	4,8	9,1	21,3	115	39,1	14,1
Cherkasy	3,5	2,9	1,9	128	1,5	1,3

Source: own calculations based on data from the State Statistics Committee of Ukraine and the Ministry of Economy of Ukraine.

declined dramatically throughout the 1990s, as the loss of long-standing links with the republics of the former Soviet Union forced local factories which had become less competitive to reduce capacities and to run up considerable debt, or even to become insolvent. Over the last three years, however, the situation has begun to improve and investment flows have increased significantly. In 2003, the machinery sector had the highest growth rate (36 per cent) with an expanding share in exports (15.8 per cent), indicating possible gradual improvement.

Ukraine's *central region*, which includes the capital city Kyiv, receives the largest share of foreign direct investment (40 per cent). This can probably be explained by economic factors such as better infrastructure, a highly qualified labour force, an efficient administrative system and the region's 'central location'. Kyiv, though its population is only half that of the Donetsk region, the country's largest and most industrialized region, has a greater share in the national gross value added and outperforms the latter with regard to several other macroeconomic indicators.

The *Podilsky, Polisky and Carpathian regions* are predominantly agricultural, with little industrial output. They have been badly hit by the agricultural crisis of the last two years and been forced to diversify their economy, notably by developing their light industry and machinery sectors along with services. The latter, and especially tourism, have become more important during the past few years, reducing social tensions by creating new employment opportunities. *Oblasts* in these regions are gradually receiving more investment, both national and foreign, as they benefit from a cheap but motivated labour force, proximity to the enlarged EU and the existence of several free economic zones. The generally positive impact Poland's, Hungary's and Slovakia's accession to the EU had on these border *oblasts* could become more permanent if cooperation between the EU and Ukraine were to lead to increased integration in the form of economic and social harmonization, trade liberalisation and free movement of persons across common borders.

The *southern region* has a mixed agricultural and industrial economy, specialising increasingly in maritime goods and services, especially transport which accounts for over 80 per cent of all Ukrainian exports in services.[1]

To sum up, economic growth after independence prompted the rehabilitation of industrial production and a considerable rise in exports, but was limited almost entirely to the metallurgical strongholds of eastern Ukraine and had no noticeably positive impact on people's welfare, not even in the regions concerned. During the last three years, economic growth, including in the industrial sector, has spread more evenly across the country, reaching central, western and southern Ukraine (see maps 6.1 and 6.2). This has been possible thanks to a diversification of industrial output, flourishing services and higher consumption levels in the domestic market. There is clear evidence of improved household welfare and higher labour productivity. However, there is a real risk that the industrial regions of eastern Ukraine will face

---

1   Services in general have always accounted for a large part of the gross domestic product, although their share has slightly dropped (from 50 per cent to 45 per cent) during the last three years and is lower than in many Central European countries.

social unrest and severe problems in the near future because of their depreciated industrial assets, while the western region are set to become the new 'locomotives' of Ukraine's industrial growth, thus reinforcing the traditional east-west rift.

From an economic point of view, differences in attitude thus make sense. Eastern *oblasts* have extensive and strong ties with former Soviet republics, and in particular Russia. Although Ukraine as a whole now depends much less on Russian and CIS markets and has more intensive trade relations with the EU, this is not yet the case in the eastern regions where more than half of all foreign trade is with Russia. A greater number of people there identify themselves as Russians. Borders with Russia are almost non-existent and, though crossing them can at times be difficult for the locals, the problems encountered are nothing compared to those prevalent at Ukraine's western borders. Should Ukraine move to strengthen all of its borders, great attention will therefore have to be paid to the eastern and northern border *oblasts*, where locals will undoubtedly face hardships because of the curtailment of informal links with neighbouring communities in Russia and Belarus. The recent experience of western Ukraine as well as the study of the way Euroregions work might here offer useful ideas about implementing a more efficient border policy. Western Ukraine, on the contrary, is far less dependent on Russian markets and has closer links with Central European neighbours and with western Europe. Though a large number of people are still employed in agriculture, the region's economic

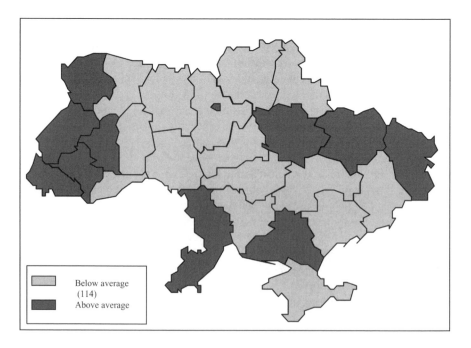

Map 6.1   Industrial growth in Ukrainian regions, 2001

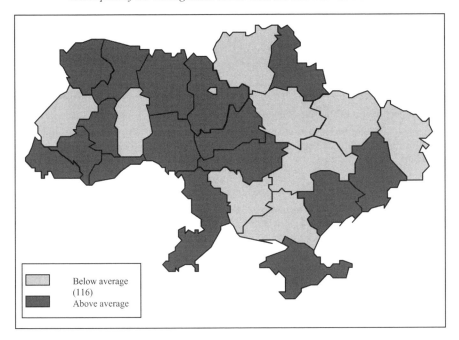

Below average
(116)

Above average

Map 6.2   Industrial growth in Ukrainian regions, 2003

activities have moved towards the production of consumer goods and the services sector. People here appear to show more economic flexibility, can look back on a tradition of private entrepreneurship and civil society, and have stronger feelings of national identity. For them, their future lies in the West.[2]

Cross-Border Cooperation as a Stimulus for Policy Harmonization

With 5,500 km of land borders, Ukraine has a great potential for developing cross-border cooperation, particularly at its highly strategic western border. The opening of this border promises to offer a high level of security to Ukraine while allowing for the free movement of goods, services, capital and people, and has therefore received top priority status from the Ukrainian government. However, it seems unlikely that this aim will be achieved in the near future as the EU, in response to new global threats to its security, is adopting a policy of tighter control. Ukraine is believed to be the biggest exporter of illegal labour and lies along major routes for

---

2   The Autonomous Republic of Crimea, which became part of the Ukrainian Soviet Socialist Republic in 1954, is mainly inhabited by Tartars, has a different historical identity due to its geopolitical location at the heart of the Europe-Russia-Asia triangle, and maintains close links with both Russia and Turkey.

smuggling human beings and illicit goods, especially arms and drugs. Therefore, the EU is paying particular attention to this border and each year spends vast sums of its budget on making it 'safer' and less permeable. This is in stark contrast to the borders with Russia and Belarus to the east and north, whose openness is one of the reasons why Ukraine has become a transit state for illegal trafficking. Achieving more open borders with the EU such appears to come at the price of a trade-off, requiring Ukraine to tighten control on its other borders.

Ukraine's neighbours to the west, who recently joined the EU, claim that 'old' EU members tend to exaggerate problems at the Ukrainian border and that control procedures there are actually quite efficient. They are therefore calling for more relaxed border formalities, especially for those Ukrainians living in the border areas as well as for businessmen, students, scholars and other categories of persons, on whom steady growth on both sides of the border is held to rely. Several of these countries have introduced 'asymmetrical' visa regimes for Ukrainian citizens: both Poland and Hungary issue visas for free, while Slovakia and the Czech Republic will be issuing free visas during a four-months trial period (from 1 May to 1 September 2005) in response to a Ukrainian initiative abolishing visas for EU citizens during the same period. This, in fact, represents a considerable breakthrough in Ukrainian foreign policy after many years of debate between pro- and anti-European forces about the possible opportunities and dangers of open borders for EU citizens. It is expected that the trial period will provide detailed information about the costs and benefits of such a policy.

Furthermore, Ukraine already fulfils all the legal requirements for developing and implementing local initiatives of CBC with EU countries. It has ratified the European Charter on Local Self-Government and the European Outline Convention on Transfrontier Cooperation between Territorial Communities or Authorities (Madrid Convention of 1980). Provisions have also been made in the Concept of State Regional Policy (2001) for the 'stimulation of cross-border and interregional economic relations, and the development of legislative proposals to broaden the authority of local and regional governments in the field of cross-border and interregional economic cooperation'. In addition, the Law on Cross-Border and Transregional Cooperation, the Draft Law on Stimulation of Regional Development (Ministry of Economy of Ukraine (2003) and the Draft National Strategy for Regional Development 2005–2015 (Ministry of Economy of Ukraine 2004) all envisage a strengthened role and greater independence for local governments with regard to cross-border cooperation. The latter document thus lists as one of its five main priorities the initiation of a process intended to bring socioeconomic procedures and standards in Ukraine's western border regions, and especially in the Euroregions, into harmony with EU legislation as a preliminary for their adoption by the state.

Formally, Ukraine's constitution, the Law of Self-Government in Ukraine (1997) and the Law on Local State Administration (1999) contain provisions defining the scope of local governments' responsibilities for local economic development and CBC, though there remain some inconsistencies between the various laws and normative documents. In reality, weak local government, its subordination to local agencies of the state administration and its lack of financial resources and capacity all mean that there is little local initiative.

Recently introduced draft legislation is therefore aiming to further reform Ukraine's territorial and administrative structure by allowing for a considerable redistribution of powers in favour of local councils, including plans to abolish state administration at *rayon* (county) level and to substantially limit its authority at *oblast* level. This would provide local governments with greater opportunities to implement projects of socioeconomic development, including within the framework of CBC, and give them greater freedom to decide which projects to fund at which level and from which budget. Currently, local authorities make virtually no contribution to internationally funded CBC projects. As a result, local and regional actors rarely consider CBC projects as serious catalysts for socioeconomic development, improved living standards or local regeneration.

Another serious problem identified by Ukraine and its neighbours in central and eastern Europe is the lack of coordination between the EU's PHARE and TACIS programmes. Thus, it is very difficult for projects on the Ukrainian-Polish border to achieve tangible and sustained results when both parties follow different procedures with different budget sources and different schedules. While the Carpathian, Bug, Upper Prut and Lower Danube Euroregions on Ukraine's western borders are probably well located to implement CBC projects, these are in danger of failing because of the limited scope of local governments' responsibilities and budgets.

The European Neighbourhood Policy (ENP) promises to remedy these difficulties. Its principles, its geographical scope and a methodology for its implementation are set out in the European Commission's Communication on Wider Europe (2003) and the subsequent ENP Strategy Paper (2004). However, it provides only universal principles and does not take into account disparate development paths and patterns among the EU's neighbours and contrasting aspirations with regard to future relations with the Union. Ukraine is currently implementing the EU-Ukraine Action Plan for 2005–2007 (Commission of the European Communities 2004c) at an accelerated rate in order to initiate negotiations about its future status as an associated member as early as 2007. Whether it will succeed or not will of course depend largely on political decisions made by the EU's current members and on public opinion in these countries.[3]

---

3   In March 2005, TSN Sofres organised an opinion poll in the six biggest EU member states (France, Germany, UK, Italy, Spain and Poland), asking 6,000 respondents about the EU's future and the countries they think should become future members. As reported by the daily *Korrespondent*, the survey found that 55 per cent of the Europeans want so see Ukraine join the EU, compared to 45 per cent who support the idea of Turkey's future membership. Unsurprisingly, Ukraine's greatest 'fans' (with 77 per cent of favourable responses) are the Poles, who have traditionally supported Ukraine and advocated its future membership. The French and the British are also supportive (with 58 per cent and 49 per cent of positive reactions respectively), while the Germans are more likely to come out 'against' (53 per cent) than 'in favour' (41 per cent) of Ukraine's membership. Those who responded favourably to the idea of a nation of 48 million people joining the club believe that Ukraine belongs to

Ukraine fully understands the legal and procedural constraints which do not allow border areas of non-members to be treated in the same way as those of accession countries, that is, that the EU's internal budget cannot be spent beyond its borders. Ukraine therefore welcomes the European Commission's recent proposal for future cross-border cooperation, especially the single-instrument approach which harmonizes budget sources and programming procedures. Until now, the TACIS Cross-border Programme has mainly focussed on the development of border infrastructure (border crossings), environmental protection and the management of natural resources, support of the private sector and assistance for economic development. Ukraine's national strategy for regional development in its border areas is already very much in line with the priorities defined by the EU's new CBC policy, which will put increased emphasis on labour markets, competitiveness, education, science, 'people-to-people' contacts and on economic policy in general. It will surely benefit from such external support, and in particular from the plan to substitute the present TACIS programme with the European Neighbourhood and Partnership Instrument (EPNI) by 2007. According to the Financial Framework for 2007–2013 (200?), countries that come under the ENP will receive € 15 billion in total, twice as much as during the previous programming period (2000–2006), to reform CBC instruments, ensuring enhanced cooperation and the implementation of EU best practice.

Despite this favourable outlook, much remains to be done at the national and local level in Ukraine. Above all, it will be necessary to raise the awareness of local communities and public bodies about issues of cross-border cooperation. Spontaneous links have existed for many years and brought benefits to communities on both sides of the frontier. But EU enlargement in 2004 has changed the Union's border requirements, and local governments now must implement new forms of interaction, organizing them properly and giving them official status. Furthermore, alongside special ENP information centres, they will have a crucial role to play in encouraging representatives of local governments, NGOs, business organizations and scientific and educational institutions to initiate CBC projects with their foreign counterparts.

Other possible initiatives include networking and twinning schemes, such as the TACIS City Twinning facility, which are among the most effective ways of transferring knowledge gained through experience. Twinning could thus take place between communities in Ukraine and in both 'old' and 'new' member states, as has already happened in the Carpathian Euroregion, and would be of particular interest to cities dealing with similar problems, for instance, Ukrainian, Silesian and Limburg coal-mining towns that have to tackle the economic and social consequences of mine closures.

Labour markets are another crucial issue. High levels of unemployment and people's relative mobility have transformed Ukraine's western border regions into

---

Europe geographically and that its aspirations cannot be denied, while Turkey is held not to be part of the European continent and not to share all European values.

one of the largest labour reservoirs for the EU. Various sectors in several EU countries depend on cheap qualified or unqualified Ukrainian labour: builders in the Czech Republic, the UK and Poland; teachers in Poland; nannies and babysitters in Italy and Greece; agricultural workers in Portugal, Spain and Greece, etc. It is estimated that some two to five million Ukrainians are working abroad and that several million Ukrainian households depend on remittances sent from there. However, most of these migrant workers have no recognized legal status and find themselves in a highly vulnerable position, without protection from dishonest employers or criminal intermediaries and living under constant threat of deportation. CBC activities dealing with labour shortages in neighbouring border regions within a legal framework could contribute much to reduce illegal labour migration.

Finally, local governments should be able to give better support to small and medium-sized enterprises, identify obstacles to the development of the private sector and learn how to overcome them. Although there exist already a fair number of projects in several *oblasts* of western Ukraine, none has produced a multiplication effect so far. Little local initiative means that local development projects still do not receive enough feedback.

United We Stand, Divided We Fall?

EU enlargement has brought about many socioeconomic and political changes in Ukraine and will continue to do so, especially in the western border regions. During the decade which followed the collapse of the Soviet bloc, the inhabitants of border regions in Ukraine, Poland, Slovakia, Hungary and Romania enjoyed substantial freedom of movement across state borders, a factor that contributed to mitigate the worst effects of economic crises. Now EU accession has imposed many obligations and rules which hinder cooperation between Ukraine and its 'old friends'. Under the new circumstances, several factors are likely to bear influence on Ukraine's economic and political future.

Poland's and Hungary's eastern border regions, just as their Ukrainian neighbours, are all on the economic periphery of their country, lack genuine economic potential and have a high negative migratory balance. They count as the poorest in the EU-25, and it remains uncertain whether the considerable structural funds allocated to them in future will in the medium term lead to significant economic development from which Ukraine might benefit, too.

In the short run, EU accession has had some rather dire consequences for the new members. By adopting single European principles for certain of their economic activities, these have become less competitive due to rising business costs. Regions in western Ukraine, in particular Zakarpatska, Lviv and Volynska *oblasts*, have, on the contrary, been experiencing an investment boom as early as 2002 and 2003. Many companies from both western Europe and the accession states have started to relocate their business to Ukrainian regions close to the EU border. Among the decisive factors for these moves have been lower labour costs, a well-developed transportation network and Ukraine's positive macroeconomic dynamic over the last four years, despite a complex and ineffective regulatory environment. Moreover, recent political events, and in particular the 'orange revolution', have ensured a more positive image of the country abroad and contributed to the creation of a more investor-friendly climate.

Many experts consider the localisation of productive capacities in western Ukraine a good starting point for expansion into the domestic market. Though labour costs have been rising slowly in recent years, it is thought that foreign investors will continue to be attracted to the region. Tax holidays for periods of 20 to 30 years, for instance, have been designed with the aim to give them the opportunity to expand and to gain a relative advantage compared to companies importing goods and services. The astonishing pace of development in some western *oblasts*, notably Zakarpatska and Lviv, can to a large extent be attributed to the existence of free economic zones.[4] However, the recent change of national leadership has led to a considerable review of economic policies and threatens to have dramatic effects: in the winter of 2005, all privileges for particular sectors or territories have thus been abolished, and discussions are under way to transform all of Ukraine into a free economic zone. This has come as quite a shock for investors, some of whom are going to court to protect their rights, while others have discontinued their projects.

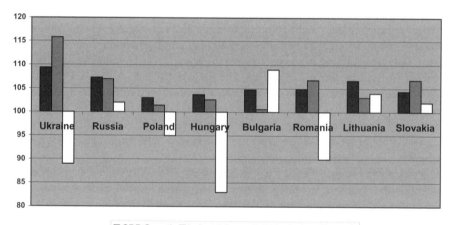

■GDP Growth ■Industrial growth □Agricultural growth

Figure 6.1   Relative macroeconomic indicators of Ukraine and some neighbouring countries in central and eastern Europe

---

4   By 1 January 2005, the EU-25 have accounted for 55 per cent of the cumulative flow (or US$ 8.3 billion) of foreign direct investment (FDI) to Ukraine. With $ 177 of FDI per capita, this is still way below the average observed for transition states in Central Europe. Central European investments tend to be concentrated in *oblasts* close to the EU border, that is Lviv (where roughly 30 per cent of FDI is of Polish origin), Zakarpatska, Volynska and Ivano-Frankivsk.

The location of European or international companies in western Ukraine will certainly help improve management culture and spread Western business standards. With proper supporting institutions, the EU neighbourhood might thus become a catalyst for flourishing private entrepreneurship. However, unless these changes are brought to bear on other parts of the country, EU enlargement risks to further entrench the rift between western and eastern Ukraine. With this in mind, the central government's regional policy should emphasize the necessity of building strong partnerships between regions in these two parts of the country in addition to the intensification of CBC activities in the western part. These partnerships, of which some already exist, such as the partnership between the Lviv and Donetsk *oblasts*, would facilitate the exchange of information about local experience and the spreading of best practice, that is, economic success in the east and the development of services and private enterprise in the west. Indeed, sound socioeconomic policies must ensure that the positive effects of these transformations will benefit the whole country. In this sense, western and central Ukraine will not be able to win the economic game without support from eastern and southern regions, and vice versa. But much will also depend on the EU: should the latter fail to pursue its open-door policy, Ukraine's still fragile European hopes might well be replaced by support for integration with Russia.

Chapter 7

Regional Cooperation in the Ukrainian-Russian Borderlands: Wider Europe or Post-Soviet Integration?

Tatiana Zhurzhenko

A major focus of this volume is the possible impact of EU enlargement on regional cooperation beyond the EU's external borders. In this essay, I will discuss both the situation in the Ukrainian-Russian borderlands and perspectives of cross-border cooperation between the two countries. The central question here is whether the EU's New Neighbourhood Policy can be seen as a geopolitical project in competition with Russia's ambitions for regional integration in the European Near Abroad. As the elections of 2004 aptly demonstrated, Ukraine is torn between two very different integration projects – one leaning eastwards, the other with a 'western' agenda. Will this competition have beneficial effects for the Ukrainian-Russian border region? Will this situation change with the institutionalization of Wider Europe and Ukraine's new status as a – theoretically – privileged 'neighbour' and in which ways? For the time being, the EU's declared aim with regard to Ukraine's eastern border is to prevent smuggling and illegal border crossing (transit migration in particular) by improving the efficiency of border controls. Thus, security looms large and it remains an open question whether the enlarged EU will be interested not only in security issues but also in supporting economic and humanitarian cooperation across this border. Using Ukrainian-Russian Euroregions as a case in point, the following will focus on perspectives and limits of cooperation that are emerging from overlapping processes of European and post-Soviet integration.

The Changing Geopolitics of the Ukrainian-Russian Border

Unlike other post-Soviet borders which, since the end of the 1980s, have become areas riven by ethnic tensions and military conflict, the new Ukrainian-Russian border has rarely received international attention. More than ten years after the two former Soviet republics 'separated', people still are able to cross the border with internal passports[1] and the prospect of a visa regime seems rather unlikely. Positive

---

1   The obligatory requirement of having an international passport for persons crossing Ukrainian-Russian border is scheduled for 2005.

memories of the recent common past, the absence of ethnic and religious conflicts as well as persistent and intensive social and cultural contacts across the border convey an impression of continuity rather than of disruption and radical change. At the same time the process of border building between Ukraine and the Russian Federation is being profoundly shaped by the momentous geopolitical transformations which have taken place in eastern Europe since 1990, that is, the collapse of the USSR and of state socialism, on the one hand, and the ambitious project of EU enlargement, on the other.

The disintegration of the Soviet Union and the declaration of national independence by the Ukrainian parliament transformed the former administrative line between two Soviet republics into an international border. Ukraine and the Russian Federation both officially recognized the boundary and declared their intention to base future relations on the principle of national sovereignty. In fact, the fundamental assumptions behind these declarations were quite different: Ukraine saw the recognition of the border as a necessary preliminary to independent statehood whereas Russia maintained that 'agreement to the borders was implicitly tied to Ukraine's full integration in the Commonwealth of Independent States (CIS) and her 'friendly attitude toward Russia' (Trenin 2001, p. 165). This divergence gave rise to serious tensions when it came to the status of Sevastopol and the Black Sea Fleet, and the nationalist and communist opposition in Russia aggressively contested what they considered the transfer of Crimea to Ukraine. Only in 1997 did Russia officially recognize the territorial integrity of independent Ukraine within its present borders through the Treaty on Cooperation, Friendship and Partnership. This treaty, which took the Russian parliament almost two years to ratify, created a basis for normal relations between the two countries, and the recently elected Russian president, Vladimir Putin, primarily for economic reasons, then adopted a more pragmatic policy towards Ukraine. During the same period the government of Ukraine's president Leonid Kuchma had become increasingly isolated from the West through its refusal to implement democratic reforms. As a consequence the government's policies shifted towards a more pro-Russian stance and thus became rather vulnerable to political pressure from Moscow. However, under these conditions some progress was achieved on the issue of the Ukrainian-Russian border. The Agreement on the State Border between Ukraine and Russia was signed in January 2003 by Putin and Kuchma and ratified by both parliaments in April of that year. The agreement finalized the negotiations on the delimitation of the Ukrainian-Russian border, or more exactly its terrestrial part – a process which had taken some four years. However, the controversial issue of the delimitation of the Azov Sea and the Kerch Strait unexpectedly caused a dangerous crisis in Ukrainian-Russian relations in October 2003 and has still not been settled. Perhaps more important, although less known to the public, is the problem of the demarcation of the terrestrial border, where a consensus between the official positions of both countries seems to be almost impossible to achieve.

The issue of demarcation illustrates quite well how EU enlargement is shaping the geopolitical context of the Ukrainian-Russian border. Ukraine insists on the demarcation of its border with Russia by referring to the principle that all state borders should have equal status. An non-demarcated and hence 'unprotected' border with Russia, often presented in public discourse as a 'dangerous openness to the East', has a symbolic meaning for the Ukrainian political elite; in their eyes, it reflects the incomplete project of state-building. Moreover, a demarcation of Ukraine's state borders according to international standards would serve as proof of the country's European choice and readiness for integration into the European Union. Russia, on the other hand, always considered borders inside the CIS to be 'internal' borders and declined to discuss demarcation on the grounds that this would be incompatible with the existing 'partnership relations' between Ukraine and Russia. The transparency of a common open border (as well as a common jurisdiction over issues of defence policy and national security) is seen as a substantial part of this partnership. Thus, whereas Ukraine strives for 'normal borders' as an attribute of the nation-state and a condition of future EU membership, Russia claims regional leadership and a place at the centre of economic integration, and therefore is more concerned about the situation in its 'borderlands' and about its geopolitical influence in the Near Abroad.

Similarly, cross-border and regional cooperation between Ukraine and Russia should be considered within the broader geopolitical context, namely its compatibility with Ukraine's ambition to become a member of the EU. The idea of cross-border cooperation was borrowed from the European context with the aim to use it as a universal instrument for solving social problems of the new Ukrainian-Russian border regions and for maintaining traditionally close relations between neighbours. Representatives of business elites from eastern Ukraine, who lobbied for this project, hoped to use the instruments of cross-border cooperation for their own economic interests which were closely linked to Russian markets and Russian sources of energy. A Council of the Border Regions of Russia and Ukraine was created in 1994 and an Agreement on the Cooperation of the Border Regions signed in 1995 by the two governments. Since 2000 and the political rapprochement between Ukraine and Russia, the idea of cross-border cooperation has received a new lease of life. Economic forums and other meetings of members of Ukrainian and Russian business circles with parliamentarians and officials from both countries have become regular events; most of them take place in Kharkiv and Belgorod. In February 2002, during a meeting of the Russian and Ukrainian presidents in Dnipropetrivsk, both parties signed a Programme of Interregional and Cross-Border Cooperation (2001–2007), in which the idea of Euroregions in the Ukrainian-Russian borderlands received official support. With the support of the Council of the Border Regions, various projects were initiated to develop 'near border' infrastructure (transport routes, border crossing points), to make common use of water resources and to protect the Siversky-Donets River, as well as to implement an experiment on encouraging cross-border trade (Kolossov and Kiryukhin 2001). In order to facilitate cooperation in education and research and to provide broader opportunities for students in both countries, the Consortium of Near-Border Ukrainian and Russian Universities was created in 2004. As early as 2003, representatives of the Kharkiv and Belgorod administrations had signed an agreement to create the Euroregion Slobozhanshchyna – the first of its kind on the Ukrainian-Russian border.

However, some hidden ambiguities persisted in these otherwise positive developments. Ukrainian politicians and officials saw cross-border cooperation with Russia rather pragmatically and widely used the rhetoric of European integration. Their Russian partners, on the other hand, tended to consider cross-border cooperation as a point of departure for more ambitious projects of economic (and political) integration and preferred to use the language of 'common historical roots' and Slavic brotherhood. It thus appears that Russians and Ukrainians disagreed on the strategic goals of cross-border cooperation, but never openly discussed these disagreements. In principle, the idea of cross-border cooperation as it was developed in the European context is, however, inseparable from other elements of supranational integration, such as the harmonization of legislation, free trade, and common security and defence policies. As a result, the Ukrainian political leadership, which officially declared EU membership a strategic goal, has not been particularly eager to move further along the road of economic and political integration with Russia, even though some concessions were made to Moscow.

Until the presidential elections of 2004 and the 'orange revolution', the principles and means of cooperation with Russia were a constant source of tension among Ukraine's ruling elites. In autumn 2003, for instance, the debate in the Ukrainian parliament on the Single Economic Area agreement (SEA) with Russia and Belarus, signed in September 2003 and ratified by the Ukrainian parliament in April 2004, thus helped consolidate the political opposition to Kuchma's regime. The agreement contained ambitious plans for regional economic integration by promising to gradually create a customs union and a free trade zone in the region, as well as to ensure harmonization of foreign trade policies, of tax regulations and, to some extent, of monetary and credit policies. If fully implemented, it would, of course, have changed the economic role of the Ukrainian-Russian border and the basic conditions for cross-border cooperation. In fact, even then, this perspective never seemed realistic (see Khokhotva 2003). The agreement was indeed heavily criticized by the political opposition in Ukraine as a betrayal of the pro-European choice, as being both anti-constitutional and economically unacceptable. Three leading ministers of the Ukrainian government openly opposed the signature of the treaty, and these reservations were reflected in the final text of the document which stated that Ukraine would respect the agreement only if it did not violate the Ukrainian constitution. Since the opposition's victory in 2004 and the advent of a new government, a positive outcome has appeared even less likely. While Ukraine is still interested in some aspects of the agreement, such as the free trade zone, it does not accept the idea of a customs union and a central body controlled by Moscow. Similarly, in 2002, Ukraine preferred 'observer status' to membership in Eurasian Economic Community (EvrAzES), the custom and economic union established under the leadership of Russia, so as not to run counter to its strategic course towards European integration.

The campaign for the presidential elections in 2004 clearly demonstrated that issues of foreign policy and geopolitical identity are at stake for the Ukrainian political class. While the programmes of the two main candidates – the opposition leader Yushchenko and his opponent, Prime Minister Yanukovich – were rather similar (as well as highly populist) in terms of social and economic development, they differed drastically in the area of foreign policy. Yushchenko, widely seen as a pro-Western candidate, called for closer integration of Ukraine into the EU and promised 'real' political and economic reforms to accelerate this process. Yanukovich stood against integration in the EU 'at all costs' and rejected negotiations where Ukraine would be in the position of a 'younger brother'. Instead, he argued for closer economic and political cooperation with Russia.[2] Additionally, he promised to recognize Russian as a second official language, to introduce dual citizenship and to facilitate cross-border contacts with Russia. In what came to be called one of the most competitive election campaigns in the post-Soviet space, a non-communist pro-Western and EU-oriented opposition became for the first time a significant and independent political force in the history of independent Ukraine.

It now appears that, despite a certain political rapprochement, increasingly diverging geopolitical orientations of Ukrainian and Russian political elites during the Kuchma era were imposing considerable constraints on cross-border cooperation. Before the wave of 'colourful revolutions' swept over the CIS, Russia tried to regain its status as a 'great power' *(velikaya derzhava)* by 'gathering together the lands' of its Near Abroad. Although disappointed by its exclusion from the European integration processes and remaining vulnerable to pressure from Moscow, Ukraine nevertheless continued to pursue a European strategy but had to take into account Russian ambitions in the region while being wary of too close integration with an economically dominant and politically unpredictable partner. Moreover, Ukrainian foreign policy under Kuchma, unlike that of Russia where the political and business class is rather consolidated as far as 'national interests' are concerned, was often manipulated by oligarchic groups and therefore unpredictable (see Bukkvoll 2004). Finally, the very character of Ukrainian oligarchic capitalism made the country's elite dependent on Moscow and afraid of Western influence when it came to such issues as democracy and human rights.

To sum up, with the EU enlargement in May 2004, Ukraine has found itself squarely between the Elephant and the Bear – between the 'reluctant empire' of the European Union and the 'reluctant ex-empire' of Russia (see Emerson 2001). As a result the Ukrainian-Russian borderlands are located at the crossroads of two competing regional integration processes: an enlarged EU seeking to institutionalize its notion of a New Neighbourhood, on the one hand, and Russia trying to re-integrate former Soviet republics under its leadership, on the other. As we will show in the last

---

2   Some observers believe that 'integration with Russia' is only a populist rhetoric of Yanukovich, who represents the interests of business clans interested in economic protectionism both from the West and from Russia (see Taras Kuzio, 'Ukraine's new pragmatic nationalism', *Kyiv Weekly*, 1 October 2004).

section of this chapter, the 'orange revolution' did not solve this dilemma. It did not change the EU's vision of Ukraine as a 'neighbour' rather than as a future member, nor is Russia prepared to renounce its interests in Ukraine. For the time being the country is thus bound to exist in the overlapping borderlands of the EU and Russia.

The Ukrainian-Russian border in the New Political Architecture of Europe

As Olga Mrinska has shown in the previous chapter, EU enlargement has significantly affected the political status of and the border regime at work on Ukraine's western frontier. My task here is to consider the enlargement's impact on the Ukrainian-Russian border, where changes might be less obvious but remain nonetheless significant. My main argument here is that the EU's strategy to strengthen its external borders implicitly encourages Ukraine to strengthen the barrier function and symbolic role of its border with Russia.

The 'imaginative geographers' behind the recent EU enlargement – politicians, bureaucrats and intellectuals – were very slow and reluctant to recognize Ukraine as a European country, rather considering it as a buffer state undeniably belonging to the Russian sphere of influence, a grey zone of the Near Abroad. Long before the enlargement process had started, the EU preferred to define its relationship with Ukraine mainly in terms of security. Support for market transformation and democratic reforms in Ukraine (rather limited in comparison to other post-communist countries) was designed more with the aim to maintain political stability in the region than to help Ukraine prepare for eventual accession. This has been admitted by experts, who see EU policies toward Ukraine, Belarus and Moldova as targeted at protecting the EU rather than at engaging with these neighbouring states (O'Rourke 2003). Due to its large migration potential, the impoverishment of its population with the concomitant risk of the spread of transmissible diseases such as tuberculosis and AIDS, its position as a transit country and its environmental problems, Ukraine is perceived especially in terms of a 'barrier' state. The EU's Country Strategy Paper 2002–2006 on Ukraine (European Commission 2001c) stresses that 'enlargement is bound to make the EU more sensitive to *"soft" security* threats from Ukraine which need to be addressed: environment (nuclear safety and related issues, including follow-up to the closure of Chernobyl and safety of Ukraine's nuclear power plants; climate change and Kyoto protocol; Black Sea protection); Justice and Home Affairs (judicial reform and combating organised crime, corruption and illegal migration); public health (transmissible diseases)' (p. 17 ; emphasis in the original text). Among the various risks associated with Ukraine as a neighbouring state, the issue of illegal migration is considered to be one of the most urgent. Ukraine is in fact the biggest transit country for migrants from the Middle East, China and NIS countries moving to western Europe. It is thus hardly surprising that Ukraine's borders are of primary interest to the EU.

As a result cooperation in Justice and Home Affairs between Ukraine and the EU has developed quite successfully, with border management as one of its focal points (see the European Council's Common Strategy on Ukraine, 1999). In addition, the Country Strategy Paper 2002–2006 and National Indicative Programme 2002–2003, with a total budget of € 115 million, have conveyed top priority to the support of institutional, legal and administrative reforms. In total, € 59 million were being allocated to this policy area, of which 22 million were earmarked for border management. The subsequent National Indicative Programme 2004–2006 (European Commission, 2003c) has a total budget of € 212 millions, 110 millions of which are allocated to institutional reforms, including 60 million for border management. It mentions that 'combating organized crime, drugs, terrorism, trafficking in human beings, support to Ukraine's efforts in strengthening its overall border management system including the eastern and southern border as well as increased cooperation in the fields of refugee protection, good governance and the judiciary are some of the common challenges' (p. 10). Most attention is being paid to Ukraine's western border; however, the EU also is concerned with improving overall border management and this includes the Ukrainian-Russian border.

The EU's aim with regard to Ukraine's eastern border is to monitor illegal migration, and transit migration in particular, and to reduce migration flows by improving the efficiency of border controls through training personnel and providing modern technical equipment. The EU Action Plan on Justice and Home Affairs concerning Ukraine (European Commission 2003d) identifies as one of the main areas of cooperation the 'development of a system of efficient, comprehensive border management (that is border control and border surveillance) on all Ukrainian borders and examination of possible participation of the State border service in a system of early prevention of illegal migration' (p. 2). For these purposes, financial and expert assistance is being provided by the EU as well as by some member states (Germany, Austria and, more recently, Poland).[3] The EU has also welcomed an initiative by Ukraine to reduce the number of border guards on its western border in order to strengthen control on the borders with Russia. With improved management of Ukraine's western borders, an increasing share of the EU's programmes will be directed to the eastern borders. The EU thus decided to finance the technical modernization of the Sumy border guard division, who control one of the longest and busiest sections of the Ukrainian-Russian border, to the amount of € 1,35 million and has already equipped the division with vehicles, radio stations, computers and an IT system for passport control.[4] As reported in the official magazine of the Ukrainian State Border Service, *Kordon*, future EU projects aim at improving border

---

3   Officially, Kyiv underlines the importance of EU assistance and 'hopes for an increase in its volume and for the implementation of such projects on the Ukrainian-Russian, Ukrainian-Belorusian and Ukrainian-Moldavian borders' (Ukraine–European Union. Official Site of the Mission of Ukraine to European Communities, <ukraine-eu.mfa.gov.ua>).

4   'Vid nadiynosti Ukrainskyh kordoniv Jevropa bude tilky u vygrashi', *Kordon*, no. 4, 2004, p. 10–11.

management in other sections of the Ukrainian-Russian border (Chernihiv, Kharkiv, Lugansk, Donetsk).[5]

The EU has also successfully encouraged Ukraine to harmonize immigration laws with those of the EU and to improve the country's capacities to deal with illegal migrants and asylum seekers. In the years 2001 to 2004, Ukraine has adopted new legislation on migration and asylum in conformity with EU standards and has allocated some € 1 million of its national budget for the construction of refugee centres, with plans to build more of them along the eastern border, near Kharkiv. Furthermore, Ukraine's State Border Service has started to use the data exchange system Arkan, designed to monitor the cross-border movement of persons and goods, and the State Committee for Migration and Nationalities has launched the information system Refugee (including a fingerprint information data base); both systems will allow better information exchange and cooperation between Ukraine's law enforcement agencies and Europol. In addition, since 2002, the EU and Ukraine have been negotiating a readmission agreement. The EU encourages similar negotiations between Ukraine and Russia, and between Ukraine and Belarus.

Several other international organizations are active at the Ukrainian-Russian border: 'Belarus, Ukraine, Moldova – against drugs', a United Nations project provides Ukrainian border guards with technical equipment; the International Organization on Migration (IMO) made available surveillance and communications equipment worth US$ 210,000 to the Kharkiv border guard division in 1998–2000 and launched a similar project for the Belarusian-Ukrainian border with a budget of US$ 300,000.[6] More recently, at a September 2004 meeting with counterparts from the Baltic States, the Austrian minister of internal affairs suggested refugee camps for Chechens might be set up in Ukraine to assist the EU in dealing with the problem. As a result of these developments, and despite the fact that EU membership seems at best a distant prospect, Ukraine is *de facto* an integral element within the formation of a comprehensive European security system. If, in future, the state borders of Ukraine become an important element of this system, the Ukrainian-Russian border will certainly adopt some features of the EU's Schengen boundaries.

In light of all this, what can be said about the future of cross-border cooperation between Ukraine and Russia? The project of a Wider Europe, which grants Ukraine the new status of a 'neighbour' and 'privileged partner', will have only a secondary impact on the Ukrainian-Russian borderlands. It is the western regions of Ukraine which might indeed profit from their new geopolitical status after EU enlargement. The New Neighbourhood Instrument, for example, has been designed to smoothen

---

5   Ibid.

6   'Jevropeyskyi vector prykordonnogo spivrobitnytstva', *Kordon*, no. 5, 2004, p. 10–11. As reported in the Ukrainian media, computer registration of all persons crossing the border at the Kharkiv-Belgorod section was implemented by April 2003, with the support of IMO who provided technical equipment and expert assistance ('na rosiysko-ukrainskomu kordoni vyprobouiutkomp'uternyimetodvylovunelegaliv', <ukr.forua.com/news/2002/04/19/144651. html>).

the negative consequences of strengthening EU external borders by supporting cross-border cooperation (see also Olga Mrinska in this book). The Ukrainian-Russian border regions are, however, not included in the list of border regions eligible for financial support, despite the fact that the New Neighbourhood Programme also mentions the importance of cooperation between the new 'neighbours'.

Nevertheless, it is likely that the EU will show an increasing interest in a secure Ukrainian-Russian border. Should the pressure of illegal migration and public concern about it continue to grow, especially as a consequence of further destabilization in the Caucasus region, Brussels will most certainly redouble efforts to promote border management in Ukraine, and at the Ukrainian-Russian border in particular. Increasing drug traffic from Afghanistan will be another reason. Under these conditions, a 'Europeanization' of the Ukrainian-Russian border means, first of all, adopting European norms of border management. The New Neighbourhood Programme might provide additional means for this purpose, insofar as it offers certain privileges to neighbouring states, that is, access to the EU's internal market and aid programmes, and not merely obligations to help prevent potential risks for EU members. Through a judicious neighbourhood policy, it is quite possible that the EU could increase its political leverage in order to influence Ukrainian policy-making.

What is the Ukrainian response to the recent EU initiatives on Wider Europe? Though rather unenthusiastic about Ukraine's status as a 'neighbour', more pragmatic members of the country's elite nevertheless envisage advantages deriving from the New Neighbourhood Initiative, such as EU support for the politically painful processes of demarcating a border with Russia that might be controlled more easily. In an interview on terrorism, Yevhen Marchuk, then head of the National Security Council, made a strong statement to the effect that 'the border of the European Union runs along the Ukrainian-Russian border – not in the economic sense,but in the sense of security'.[7] Indeed, the EU considers the demarcation of international borders an important condition for maintaining regional security. At the same time the examples of Latvia and Estonia and their latent border conflicts with Russia demonstrate the limits of the EU's influence on Russian positions. The Ukrainian-Russian conflict over Tuzla island in October 2003 also seems to indicate that European leaders are not willing to jeopardize relations with Russia for a mere 'family quarrel' between two CIS countries.

At the level of border politics, the process of EU enlargement thus reproduces to some extent the notorious Ukrainian geopolitical alternative of whether to join Europe or to remain with Russia. Ukrainian experts point out that strengthening the border with Russia (and Belarus) is a necessary price to pay if Ukraine wants to achieve compromises on border regimes with its western neighbours. According to Anatoliy Baronin (2001), the negative consequences of the EU enlargement at Ukraine's western border could be softened by building a 'normal' border with Russia: 'Ukraine cannot have both borders open simultaneously. Ukraine will have

---

7   See Marchuk's website at <www.marchuk.kiev.ua>.

to decide between Poland and Russia.' In official rhetoric, the unpopular policy of strengthening border controls with Russia is presented as necessary in order to maintain 'civilized borders'. The idea here is that a European country like Ukraine should also have 'European borders', that is, borders that allow for 'civilized procedures' for ordinary citizens and at the same time act as an effective barrier against contraband, crime and illegal migration. The concept of 'civilized borders' has, of course, different implications for the western and eastern borders of Ukraine. On the formerly closed and impermeable western border, it means simplification and standardization of border controls; on the eastern border, there is a strong risk of limiting the still relatively free movement of people.

Cross-Border Cooperation in the Post-Soviet Context: Towards Europe?

Let us now return to the question formulated at the very beginning of this chapter: as far as cooperation projects in the Ukrainian-Russian borderlands are concerned, what might the 'overlapping' of the EU's Wider Europe project and Russian-led integration of the post-Soviet space signify? Do these two projects contradict each other, are they in competition with each other, or might their 'meeting' result in something new? More specifically, can the New Neighbourhood Programme promote a form of Ukrainian-Russian cross-border cooperation that can serve the interests of both countries as well as the EU. In other words, can it help to 'Europeanize' this cooperation in a productive and inclusive way?

In Ukraine, there are usually two ways to approach issues of Ukrainian-Russian cross-border cooperation. Pro-European discourse gives priority to cooperation with Ukraine's western neighbours seen as agents of 'Europeanization'. Cross-border cooperation with the new EU members is considered a step towards future European integration. At the same time the perspective of cross-border cooperation with Russia is either ignored or viewed with suspicion as being contaminated by the 'Asian' virus of authoritarianism and corruption and as an anti-European factor. In the pro-Russian discourse, it is common to complain about 'obstacles' to developing viable forms of cooperation between Ukraine and Russia and about the lack of political will and 'national egoism' of the postcommunist elites.

If we put the underlying ideologies into context, the arguments of both critics and supporters of cross-border cooperation with Russia are not far off the mark. It is true that the logic of nation-building and state-building processes in the post-Soviet space in both countries imposes some limits on any cross-border cooperation project. These have much to do with the danger of economic decentralization and separatism, the necessity to protect national economies against criminal elements and the challenges involved in shaping national identity and a national cultural space. The new state institutions, border and customs control services in particular, as well as commercial firms have an economic interest in maintaining the barrier function of the border. A lack of effective regional policies, coupled with weak local governments and a dearth of financial resources, imposes severe constraints on cross-border cooperation projects in both western and eastern regions of Ukraine. Regional economic actors who might be important in strengthening cooperation are heavily involved in the cross-border 'shadow economy', which ranges from semi-legal operations to trivial contraband and corruption. Despite the official rhetoric at the regional level and some symbolic initiatives, these actors do not appear genuinely interested in new legislation regarding cross-border cooperation nor do they engage in any real lobbying effort for improved cooperation. No wonder that the results of cross-border cooperation are often disappointing and limited to bureaucratic activities.

Even more than the logic of post-Soviet state-building, it is the nature of the political regimes which have developed in both Ukraine and Russia that are responsible for this state of affairs. The Ukrainian political scientist Olexandr Derhachov (2003) argued that Ukrainian-Russian relations are, to a large extent, shaped by the narrow interests of business clans and therefore 'integration projects in the region are also a result of connections between the national "parties of power" and a means of their mutual support'. In the absence of an articulated foreign policy agenda, the Ukrainian leadership during the Kuchma era preferred to sign agreements with Russia which mainly served the short-term particularistic interests of oligarchic groups on both sides. At the same time it was implicitly assumed that the implementation of such agreements would ultimately be sabotaged or undermined by the Ukrainian state bureaucracy.

Zimmer (2004) has argued that the neopatrimonial post-Soviet regime and its client capitalism have given rise to 'captured regions' in which oligarchic clans monopolize regional development activities and subjugate them to their group interests, thus blocking competition and obstructing transparent and democratic regional policies. It is no secret that in Ukraine special privileges for regions, such as the creation of a free trade zone (FTZ) or special investment regimes, have often been abused in the political trade between the centre and the regional elites and granted as a reward for the loyalty of regional leaders to the president, especially after elections. The Donetsk FTZ, for instance, was widely used to avoid taxation and was even characterized by President Kuchma as a 'semi-criminal zone'. Instead of attracting foreign investment, the Donetsk FTZ served mainly as a mechanism through which to invest laundered money returning from offshore companies. Similarly, cross-border cooperation projects have been open to political manipulation and been exploited for electoralist purposes. The rapid rail link between Kharkiv and Belgorod, which does not stop at the border for customs checks, was opened just one month ahead of the presidential elections in October 2004 in support of Yanukovich's pro-Russian election programme.

However, all this does not mean that the eastern regions of Ukraine bordering Russia radically differ from their counterparts in western Ukraine with regard to political and economic culture or in any 'civilizational' sense. Historically accumulated 'culture' indeed matters, it creates a spectrum of alternatives for the regional elites. But how these alternatives are realized in the process of 'region-building' depends very much on 'institutionalized opportunities' such as legal frameworks and, more broadly, on the national 'transformation regime' itself (see Tatur 2004, p.31). The specificity of Ukraine's postcommunist transformation does not so much favour democratic regional development as rather help elites in eastern Ukraine profit in the short term from a stronger representation in Kiev. Under these conditions, the EU's 2004 enlargement can be seen as a geopolitical opportunity for regions in western Ukraine to establish closer links with their European neighbours, perhaps including certain elements of integration with the EU. From our point of view, it is not that the elites of Lviv or Ivano-Frankivsk are more 'westernized', rather the proximity and potential institutional presence of the EU offer this region an opportunity to develop into a 'gateway to Europe'. The main problem of the regions bordering Russia is therefore not how to develop cross-border cooperation, but instead how to promote cooperation in the absence of 'Europe', in the form of EU incentives and financial resources.

Arguments put forward in favour of cooperation and economic integration between Ukraine and Russia usually include similar transformation processes in and institutional structures of both countries, traditional links between enterprises on both sides and mutual economic interests. One could argue, however, that these similarities, originating mainly from the common Soviet past, rather create a symbolic and institutional 'lock-in' effect, preserving and even reinforcing precisely those elements that require reform and modernization. This refers first of all to the political and legal conditions for business that have favoured corruption and opportunism. The same applies to human rights, the development of a local civil society and the freedom of regional media. If those political analysts who warn of a general shift of CIS countries toward authoritarianism are correct, the question is how to ward off these possible negative tendencies while developing effective cooperation.

Several practical issues are in urgent need of attention at the regional level and would be of great benefit to the population: simplified border crossing procedures for residents of border region communities as well as cooperation on environmental policy or the development of regional tourism. Small businesses, NGO's, educational and research institutions could be very active here if freed from bureaucratic control. Some projects, such as the preservation of the Siversky Donets River, which would be vital for this overpopulated and highly industrialized region, could have a significant social impact. In principle, both the Ukrainian and the Russian border regions could cooperate in creating attractive conditions for foreign investment. However, the highly bureaucratic and often populist nature of cross-border projects, the interests of business clans as well as a lack of public involvement tend to work against a more optimistic scenario.

In this sense the role of the EU as an independent player, a modernizing factor and a normative force for democratic transformation could be very important. The other question is to what extent the EU is prepared to be active not only in nearby borderlands such as, for example, the Polish-Ukrainian border region, but also in other regions of Ukraine. It would appear to lie in the interest of the EU to support Ukrainian-Russian cross-border cooperation and thus help to modernize regional policy in both countries and also to prevent an internal geopolitical split between a 'Europeanized' western Ukraine and a 'Eurasian' East.

Slobozhanshchyna – a Euroregion outside the EU?

In this last section, we will have a closer look at a new project aiming at the creation of a Euroregion (see Béla Baranyi in this book for a discussion of Euroregions). The Euroregion Slobozhanshchyna, located on the border between Ukraine and Russia, can be seen as hybrid product of post-Soviet and European integration processes. The question posed here is whether such a politically constructed region can offer perspectives for effective and constructive cooperation between the two countries and at the same time bring the Ukrainian-Russian borderlands closer to 'Europe'.

This Euroregion project was initiated by the regional administrations of the Kharkiv *oblast* (Ukraine) and the Belgorod *oblast* (Russian Federation). It is not by chance that these two regions have pioneered this initiative. Here, the cooperation context is influenced by an interesting combination of preserved cultural, linguistic and religious similarities across the border, frequent contacts between people and, at the same time, a growing economic asymmetry. Starting in the second half of the nineteenth century, Kharkiv went through an intensive capitalist modernization and became one of the largest industrial and cultural centres of the Russian Empire as well as an important transport junction and university town. Kharkiv was able to maintain this role during the Soviet period. However, most of the Kharkiv's industrial enterprises were deeply involved in all-Union economic cooperation and thus subordinated directly to Moscow, not to Kiev. Furthermore, modern research and industrial complexes for aircraft construction and the aviation industry, machine construction, nuclear physics and physics of new materials transformed Kharkiv into one of the main scientific and industrial centres of the USSR, attracting labour and university graduates from neighbouring regions, including Russia. Belgorod, on the other hand, became an *oblast* centre only after the second world war and, though it developed quite successfully, could not compete in regional importance with Kharkiv.

This situation began to change in 1991 due to the divergent paths of economic and political development in both countries. Economic revival set in earlier and with more intensity in Belgorod *oblast* than in Kharkiv (Kolossov and Vendina 2002). The reasons for this are a better balanced economic structure and the advantages of a relatively late modernization. The rather profitable extraction and processing of iron ore as well as a developed construction materials and food industry helped Belgorod *oblast* to become self-sustainable, whereas Kharkiv, with its energy-intensive industrial complex, became dependent on the supply of Russian gas and electricity. Despite the fact that Kharkiv *oblast* is one of the most industrially developed in Ukraine and that Belgorod *oblast* is rather agricultural, average salaries in the Kharkiv region are only 80 per cent of what they are in Belgorod. Correspondingly, on the Russian side, pensions and other social benefits are higher, taxes and rates are lower, and many social indicators are more favourable, such as housing provision, medical services and crime levels. According to a June 2001 survey conducted among residents of both *oblasts* to evaluate the frequency of border crossings, the level of social optimism and satisfaction with the social and economic situation in the country is also higher among Russian citizens (Kolossov and Vendina 2002). Only 5.7 per cent of the respondents from Belgorod *oblast* expressed a desire to move to Ukraine, whereas 42.6 per cent of the Kharkiv respondents were ready to settle in Russia. It is therefore not surprising that labour migration from the Kharkiv region to Belgorod exceeds flows in the opposite direction. One can agree with Kolossov and Vendina (ibid.) that, in terms of cross-border migration, the border regions of Russia are presently more attractive for Ukrainians than vice versa.

Problems to be solved by cross-border cooperation in these borderland are many. They include the development of a legal basis for cooperation between Ukrainian and Russian enterprises, the attraction of investors for common projects, infrastructure development, the improvement of tax and customs legislation for cross-border trade as well as the development of regulations for labour migration and social guarantees for foreign workers. However, it is not merely economic issues that are at stake, cooperation is also urgently required with regards to environmental protection, academic exchange and scientific cooperation, public safety and combating illegal border traffic. Last but not least, cooperation initiatives have a humanitarian aspect in that they address the needs of local residents who work or have family living on the other side of the border.

Among the projects which are most often discussed by business elites and regional administrators of both countries are: 1) the formation of a Russian-Ukrainian consortium of banks and industrial enterprises with the aim to encourage cooperation programmes and to coordinate export policies in Russia, Ukraine and third countries; 2) testing a simplified customs and border regime that meets the social and economic needs of local populations in the Kharkiv and Belgorod regions (and with prospects of implementation elsewhere); 3) harmonizing tax systems in order to eliminate obstacles to trade; and 4) an interregional ecological programme aimed at the sustainable and rational use of the water resources of the Siversky-Donets River Basin and the creation of a Russian-Ukrainian interregional ecological fund.

Both regions are interested in exploiting their logistical advantages and therefore cooperating on the improvement of border and customs controls, modernizing crossing points and regulating transport flows. On the Russian side, the new international crossing point of Nechoteevka, designed according to modern technical standards, was officially opened on 22 June 2002 by the prime ministers of both countries. One fifth of the total goods transported between the two countries goes through the Nechoteevka crossing point and the corresponding Ukrainian crossing point Hoptovka. The next project along these lines will be the Pletnyovka border station on the Ukrainian side. Nevertheless, there is still an urgent need for constructing new crossing points and modernizing existing ones. Specialists suggest that a new scheme for cross-border transportation flows should be developed.

According to some sources, there are even more ambitious plans on the horizon. For example, a 'cross-border industrial complex' is to be inaugurated between the small towns of Shebekino (Russia) and Vovchansk (Ukraine) (Klochkov and Kiryukhin 2001). This development project, endowed with special investment incentives and a high concentration of innovative small and medium-sized enterprises, could potentially specialize in the development of new technologies. The intention is to help revitalize these economically depressed border areas, to create new jobs for the highly qualified regional work force and to help decentralize industries located in urbanized areas. It is hoped that a new and innovative industrial core will provide strong impulses for the economic and social development of the wider region, including Kharkiv and Belgorod. Another project under consideration, 'the Ecoregion', will cover an even larger area (Kiryukhin 2000). In addition to these two ambitious initiatives, there are plans for cooperation projects between Russian and Ukrainian universities ('knowledge without borders') and for the development of international tourism based on cross-border cultural routes.

Both experts and regional agencies are aware that these projects and initiatives would stand a better chance of realization if they were institutionalized within the framework of a Euroregion. The idea of a Euroregion Slobozhanshchyna, which would be the first such institution on the Ukrainian-Russian border, has now been under development for several years by specialists from the Department of Geography at the Kharkiv National University. As early as 1997, the idea of a Euroregion covering the territory of the Kharkiv and Belgorod *oblasts* was discussed by Kharkiv geographers (Golikov and Chernomaz 1997), who stressed the existence of various geopolitical, economic, ecological and cultural preconditions for more intensive cooperation. Among the factors favouring a common sense of region are not only the traditional economic cooperation between Kharkiv and Belgorod, but the emergence of new transnational enterprises and the potential benefits from international transport flows. In addition, ethnic and cultural affinities, the traditional cooperation between academic institutions, similarities of the educational systems and the common informational space are seen to support a common regional platform,too. At the political level the idea of a Euroregion received support from the Council of the Border Regions of Russia and Ukraine and, in October 2003, an official agreement was signed by the representatives of the Kharkiv and the Belgorod regional administrations.

It is still too early to judge the project by its practical results and by its economic and social impact. But even more relevant in the context of this chapter is another question: what can be European about this 'Euroregion'? Formally speaking, Euroregions have no single legal status. According to the official web site of the Council of Europe, the term 'suggests simply a feeling of belonging to Europe and a willingness to participate in the process of European integration'.[8] But is a Euroregion on the Ukrainian-Russian border not inherently a logical contradiction? One can

---

8   See   <www.coe.int/T/E/Legal_Affairs/Local_and_regional_Democracy/Transfrontier_co-operation/Euroregions>.

understand sceptics who see this project as an attempt made by some regional business clans to use the fashionable rhetoric of European integration for their own purposes. We would rather suggest that it reflects an aspiration of Ukrainian regional elites to modernize the ideology of Ukrainian-Russian cooperation which has traditionally been based on a common Slavic identity and a common geopolitical fate. But to what extent are the representatives of the regional elite prepared to comply with European norms and values? Without the democratization of local governments and without the decentralization of power in both Ukraine and Russia, this Euroregion will remain largely symbolic in nature. In this sense, regional initiatives in eastern and western Ukraine face the same problems and their success will depend on the fate of democratic reforms.

Slobozhanshchyna can be seen as an early indicator of a new phenomenon. It represents a new generation of Euroregions which are about to emerge outside the EU, along the borders separating former Soviet republics. In this sense, it is also important for the future of the EU and its neighbourhood policy to understand the specific problems of these regions. Looking back, the first Euroregions in western Europe were an instrument of economic and political integration within the EU. They were also meant to help overcome the hostilities of the past and the negative images of neighbouring communities. The second generation of Euroregions established along the borders of western Europe with postcommunist countries of central Europe had a similar task, namely to prepare these countries for accession and to educate regional elites through cooperation projects and by overcoming cold war legacies in people's minds (see Gzegorz Gorzelak in this book). Additionally, economic asymmetry emerged as a major political and social challenge after the changes of 1989 and 1990 (Kruglashov 2004). The Euroregions of the third generation, which will be at the centre of future EU neighbourhood policies, are forming along the western borders between the former USSR and the enlarged EU. Here, apart from migration pressures and ecological threats, the challenges facing the EU are quite different. Economic gaps are not only significant, they are expected to increase with enlargement. The 'new neighbours' are vacillating between authoritarian temptations and democratic ambitions. They are perceived as European, but of a second-rate category. Most importantly, the EU will have to decouple cross-border cooperation from perspectives of future membership, considerably weakening the 'Europeanizing' effect of its policies.

But if Slobozhanshyna represents a paradigmatic case for the fourth generation of Euroregions, the challenges for the EU are even more formidable. Here, due to the absence of an accession perspective (and a lack of neighbourhood policy instruments), the regional elites are not very sensitive to the incentives and sanctions of the EU. The very idea of cross-border cooperation in the post-Soviet context is also rather ambivalent as the interests of national elites often do not coincide with the nostalgic mood of the citizens. The common Soviet past sustains a certain sense of 'imagined community' even today, and the inertia of the post-Soviet institutions makes the modernization of regional policy very difficult. Probably most important, the role of Europe as an integrating force is challenged here by Russia and its own competing integration projects. Some possible cooperative venture between Ukraine and Russia, such as in aircraft construction, are seen by the EU with suspicion as generating potential competition on world markets. Issues such as these explain the EU's hesitancy to develop policies and strategies that might support Euroregions along the Ukrainian-Russian border.

Conclusion

Nevertheless, from our point of view it would be a mistake for the EU to limit its neighbourhood policy only to the western regions of Ukraine. Encouraging regional and cross-border cooperation in the 'new' eastern Europe could become an important strategy of providing security and stability to the region. As shown above, effective border management and combating cross-border criminality are urgent issues for the EU and will shape its agenda on the Ukrainian-Russian border. Appearances to the contrary, this border is becoming part of a comprehensive system of European security. At the same time support for ecological, cultural and small business projects on this border could become a powerful instrument of EU neighbourhood policy. Measures such as these would help promote the transparency of regional authorities, the reform of local self-administration and democratic reform. They could also help to overcome the prevailing attitudes that consider Ukrainian-Russian and Ukrainian-European cooperation as a mutually exclusive alternative.

It is the western border regions of Ukraine which will directly benefit from cooperation with the EU within the framework of the New Neighbourhood Programme. But the eastern border regions could also benefit from a growing EU interest in Ukraine as a new neighbour, from participation in certain European structures (such as the European Association of Border Regions), and from the experiences of Euroregions on the western borders. One should not even exclude the possibility that a Ukrainian-Russian Euroregion could prove successful economically due to its industrial potential and developed infrastructure. However, the perspectives of Ukrainian-Russian cooperation in the long term depend on the relations between the Elephant and the Bear – between the EU and Russia. A competition between these 'two empires' in the wider European neighbourhood runs the risk of deepening regional divisions inside Ukraine and thus destabilizing the country's political situation. Alternatively, true rapprochement between Russia and the EU along the lines envisaged in the idea of a common European space could encourage Ukrainian-Russian cross-border cooperation and a real, not only rhetorical, 'Europeanization' of regional policy.

Chapter 8

The New Neighbourhood – a 'Constitution' for Cross-Border Cooperation?

Ilkka Liikanen and Petri Virtanen

This essay will examine the feasibility of regional institutional innovations in elaborating EU border policies and new instruments of cross-border cooperation. The focus of the study is on the external borders of the EU. Within this context, Finnish-Russian border areas, and the Euregio Karelia in particular, will be analyzed in terms of a case study of regional cooperation at pre-enlargement borders.

In the first part of the study, different EU doctrines, instruments of cross-border cooperation (CBC) and their roles in defining EU border policies will be addressed. Attention will be drawn to the EU's Wider Europe strategy and possible consequences of the New Neighbourhood Initiative (NNI) for cross-border interaction on the eastern borders of the EU. The historical and political development of the Finnish-Russian border region will also be scrutinized in order to clarify the existing preconditions for the adaptation of new CBC instruments. The main part of the study is devoted to Euroregions and their role in implementing EU border policies and focuses on the Finnish-Russian border and Euregio Karelia. In the concluding part of this essay potential roles of regional institutional innovations in adapting the EU's New Neighbourhood programme will be discussed.

The New Neighbourhood Policy Framework and Instrument from a Regional Point of View

The explanatory memorandum introducing the Commission's proposal on the European Neighbourhood and Partnership Instrument (ENPI) opens rather prosaically with a reference to the Commission's earlier communications on financial perspectives (Commission of the European Communities 2004b, p. 2). The blunt opening is, however, well justified by the fact that the necessity of creating the new instrument was, indeed, introduced before the actual policy proposal in the Commission's communication Building Our Common Future. Policy Challenges and Budgetary Means of the Enlarged Union 2007–2013 (Commission of the European

Communities 2004g) targeted at streamlining administration in the field of external relations and CBC.

In the communication on financial perspectives the need to create new policy instruments was derived from the more general principle of 'simplifying instruments to improve delivery.' According to the communication, the success of policies relies on 'the efficiency of delivery instruments and, more broadly, in the appropriate system of economic governance'. In order to maximize performance, there was a need to strengthen cooperation and partnership, and especially for *'simplified instruments, replacing the existing* variety of complex decision-making powers and policy instruments, ranging from local and regional to national and EU levels' (ibid., p. 30; emphasis in the original text).

Following this simplification principle the communication defined two future building blocks: introduction of an integration policy roadmap and the simplification of the instruments of expenditure management. Strongly highlighted in the text, the Commission committed itself to adopting the principle of *one instrument per policy area, one fund per programme*. It stated that 'EU funding instruments will as far as possible be consolidated and rationalised so that *each policy area* responsible for operational expenditure has a *single funding instrument* covering the full range of its interventions' (ibid., p. 32; emphasis in the original text).

In regard to the area of external relations, the Commission stated that a simplified architecture will be proposed, based on six instruments, replacing the present more than one hundred different instruments. More specifically the commission proposed the creation of a New Neighbourhood Instrument, saying

> It is in the interest of the whole EU to avoid the emergence of new dividing lines on and around its external borders and to promote stability and prosperity within and beyond these external borders. A joined and united effort by the EU is vital to achieve this ambitious objective. Member States are not well equipped to address this challenge on an individual basis. [...] The cross-border component of the New Neighbourhood Instrument would be implemented as simply as possible through a single legal instrument, building on the principles of existing European cross-border programmes – partnership, multi-annual programming and co-financing (ibid., p. 33).

New Neighbourhood: One Framework – One Instrument

The proposal for policy regulation submitted later draws openly on these principles. After the first two articles defining the subject matter and scope of the regulation, the third article is committed to defining the policy framework the instrument will implement. In the explanatory memorandum this is said to underline the character of the ENPI as a policy-driven instrument: 'The overall policy framework for programming assistance should be laid down, taking into account the existing agreements, in the Commission communications and Council conclusions setting out the overall strategy of the Union vis-a-vis neighbouring countries.' (Commission of the European Communities 2004b, p. 3)

The policy-driven nature of the instrument is further emphasized in Article 7, which concerns the programming and allocation of funds. The priorities for assistance and financial allocations are said to be determined by the strategy papers drawing on the policy framework described in Article 3. In this respect, the proposal faithfully follows the principle of 'one policy framework – one instrument' defined in the communication on the financial perspectives.

In supporting administrative efficiency the explanatory memorandum naturally also stresses the factual goals of the instrument and the innovative features of its cross-border cooperation component, which is said to emphasize the role and significance of the regions. In a comment on Article 1 the scope of the assistance is defined by referring to the importance of developing an area of prosperity and close cooperation involving the European Union and the neighbouring countries, 'as recognised in the Draft Constitution' (ibid., p. 2).

Even though the reference to the Constitution is limited to this one sentence, we may ask whether in the composition of the document features exist which even in a deeper sense are akin to the guiding principles behind the Constitution. This question is especially relevant in regard of the account of the work of the 'Peace Group', which was set up by the Commission to draft the first guidelines for ENPI:

> As far as external community instruments are concerned, the 'Peace Group' recognised that the European Union's co-operation and assistance policy is the result of 50 years of successive sedimentation, which results in a multiplication of assistance instruments and a fragmentation of aid management both in terms of programming and implementation functions (even if recent policy and structural reforms have helped to improve coherence and consistency of the European Union's co-operation and assistance policy).
>
> The 'Peace Group' put forward that the European Union's framework for external assistance should be rationalised and simplified by a reduction in the number of legal bases, the number of budget lines, and the number of programmes. More precisely, it recommended that:
>
> the complex structure of existing aid programs … covering a wide range of interventions … should be significantly streamlined;
>
> European Community and Member States policies and implementation should be harmonised. (Ibid., p. 39)

As we know today the desire to streamline and harmonize EU policies put forward in the Draft Constitution has since been rejected by European citizens in the constitutional referendums in France and the Netherlands, and the whole effort has come to a standstill. At the same time the creation of new administrative policy instruments has also been endangered by the budget quarrels the member states are caught up in. In this situation there is reason to ask what political rationale can be found behind the elaboration of the ENPI and to what degree the streamlining of the policy frame behind the New Neighbourhood Initiative resembles the controversial elements in the Draft Constitution.

Constitutionalizing Cross-Border Cooperation?

Does New Neighbourhood represent a similar uniform policy façade behind which there are multi-layered ambitions and interests which only seemingly form a common policy frame? At first sight it is rather easy to distinguish certain basic trends typical of different spatial scales: the interest of EU supranational administration to streamline its organization and strengthen its authority, the nation-state-bound security interests of pacifying the borders by allying and identifying with the greater European whole, and the local and regional aspirations to find compensatory mechanisms for the economic and structural weaknesses deriving from the peripheral position. The question concerns, however, not the existence of multiple policy approaches but rather the manner in which these elements can be adjusted under one policy framework and one policy instrument – or whether this ambition represents exactly the kind of constitutionalization that today has brought the process of integration to a standstill.

In this essay the streamlining of EU border policies is examined from a regional point of view. Incongruous approaches and interests on the national and international levels are left for future analysis even though the Finnish Northern Dimension initiative, and especially Russia's negative attitude towards the EU New Neighbourhood policy framework, can, indeed, be seen as major indications of the multilayered complicity of the new policy instrument.[1] Starting from the regional perspective it is, however, vital to first examine the elaboration of EU border policies and instruments during the earlier phases of EU enlargement and to ask how much room was left for regional solutions and institutional innovations in the previous policy documents. In the second part of the study the various existing EU doctrines, instruments of cross-border cooperation (CBC) and their role in the development of the EU border policies are reviewed with special emphasis on their implications on cross-border interaction on the EU's eastern borders. The main part of the study focuses on Euregions and their existing and prospective role in the EU border policies. Especially the Finnish-Russian border and Euregio Karelia are taken under more detailed examination. As a conclusion the role of regional institutional innovations in the future adaptation of the New Neighbourhood programme is discussed.

---

1   In the Commission's policy proposal the problem of Russia's refusal to adopt the New Neighbourhood frame is bypassed by stating that the EU and Russia have decided to develop their relations in the framework of a 'strategic partnership' based on four common spaces, rather than through the European Neighbourhood Policy. 'However, the issues discussed in this framework are largely similar to those dealt with in the ENP context. For this reason the ENPI will also cover Community assistance to Russia' (Commission of the European Communities 2004b, p. 38).

Formation of the Policy Framework

EU's Common Foreign and Security Policy (CFSP)

The role of external borders in European foreign and security politics has changed. Today the primary question is no longer how to stop a military threat. After the collapse of the iron curtain, ideas and metaphors of military threat have been more and more replaced by new images of soft security risks (Grabbe 2000, p. 3). The concept of security in European discussions has expanded to include conceptions such as 'broad security', 'human security' and especially 'soft security' (Sjuren 2003, p. 2). This has also set new challenges for EU foreign policy, even though security questions have remained or even expanded on its agenda.

The basis of the Common Foreign and Security Policy (CFSP) of the EU was built as the second pillar of the European Union in the 1993 Treaty on European Union signed at Maastricht. Since then a number of changes and amendments have been introduced, such as in the Amsterdam Treaty (in which fundamental objectives of CFSP were declared) and the Nice Treaty. According to the 1997 Treaty on European Union (Amsterdam Treaty)

> [t]he Union shall define and implement a common foreign and security policy covering all areas of foreign and security policy, the objectives of which shall be:
> - to safeguard the common values, fundamental interests, independence and integrity of the Union in conformity with the principles of the United Nations Charter,
> - to strengthen the security of the Union in all ways,
> - to preserve peace and strengthen international security, in accordance with the principles of the United Nations Charter, as well as the principles of the Helsinki Final Act and the objectives of the Paris Charter, including those on external borders,
> - to promote international cooperation,
> - to develop and consolidate democracy and the rule of law, and respect for human rights and fundamental freedoms. (EC Treaty 2002, p. C325, 13–14)

Though principles of the CFSP are listed in the Amsterdam Treaty and all member states are expected to support it, there are, however, conditions that reduce its power. First, CFSP is a new phenomenon and was not included in the first contracts or declarations and there were no settled budgetary arrangements before 1997. Second, the success of CFSP is based on common coordinated actions which require the political will of the member countries. Thirdly, the historical points of departure of the member countries are different and their national interests sometimes differ markedly. Though security politics seldom leave much room for regional actors to participate in their adaptation, there has been considerable space in the EU left for the member states to consider in which actions they wish to participate in. The other side of the coin has been the threat that a common foreign policy will become an extension of the foreign policy of the larger member countries (Piipponen 2003, p. 27; Laffan et al. 2000, p. 179).

Northern Dimension as a Tool of CBC on the Eastern Border of the EU

The idea of the Northern Dimension (ND) was launched at the European Council meeting in Luxembourg in 1997, and the following year the Commission described the tentative rule of the ND in its communication. The Commission stated that the Northern region is of particular significance to the European Union and also noted that this region is the Union's only direct geographical link with the Russian Federation and, as such, is important for cooperation and communication between the EU and Russia (Commission of the European Communities 1998).

In 2000 the Council of the European Union presented the Action Plan for the Northern Dimension in the external and cross-border policies of the European Union 2002–2003 (Council of the European Union 2000). This was a reference document for actions planned and implemented in the ND region and consisted of a) a *horizontal part* that recalled the major challenges associated with northern Europe, the priorities for action and the legal, institutional and financial framework for activities relating to the ND, and of b) an *operational part* that set out objectives and perspectives for actions during the period of 2002–2003 in sectors where expected added value was greatest. In 2003 the Commission published the Second Northern Dimension Action Plan 2004–2006 (Commission of the European Communities 2003e) setting the contribution of CBC to promoting regional economic development and integration of the whole ND area as one of its main priorities. It also emphasized that various CBC arrangements, including Euroregions, can form a basis for intensified work focused on the borders with Russia in the ND region. To help address this, ND partners would work to achieve key objectives such as the implementation of CBC projects on priority themes, strengthening coordination among funding programmes, developing local-level participation and commitment to CBC and addressing the key challenges faced by the Kaliningrad region (ibid., p. 11–12).

In the CBC perspective the ND includes three important elements. First, it offers a new security perspective that complements traditional military-oriented security policy by emphasizing soft security issues like environmental questions, infectious diseases and smuggling. The second point is that the ND aims to better raise the problems of the northern regions. Thirdly, it does not include the creation of a formal institution but aims to improve long-term strategy in favour of the interests of different actors in CBC (Scott 2004, p. 135–6). The lack of a formal institutional structure and specific financial Northern Dimension instruments is also one of the weaknesses of the ND. Without specifically targeted and allocated financial instruments there is a danger that the Northern Dimension will remain a symbolic dimension and not become a serious programme. This would also reduce the possibilities of regional level CBC in the ND area.

The Partnership and Cooperation Agreement and the EU's Common Strategy towards Russia

The legal basis for the relationships between the EU and Russia were defined in the Partnership and Cooperation Agreement (PCA) in 1997 (Commission of the European Communities 1997b). In 2004 the PCA was extended and a Protocol to the Partnership and Cooperation Agreement was signed to extend the agreement to the ten new member states (Commission of the European Communities 2004c). It did not include any changes in basic guidelines but added new members as parties to the agreement. In 1999 the European Union declared its Common Strategy on Russia (Council of the European Union, 1999b). In this document the EU stated two strategic aims for cooperation. The first is a stable, open and pluralistic democracy in Russia, governed by the rule of law and underpinning a prosperous market economy benefiting alike all the people of Russia and the European Union. The second is to maintain European stability, promote global security and respond to the common challenges of the continent through intensified cooperation with Russia. The PCA and the EU Common Strategy on Russia set the basic guidelines and strategies of the EU in its cooperation with the Russian Federation.

At the St. Petersburg summit in May 2003 the EU and Russia confirmed their commitment to further strengthen and develop their cooperation. This was later confirmed in the Communication on EU-Russia relations in 2004 (Commission of the European Communities 2004d). This document emphasizes earlier commitments made in the European Neighbourhood Policy, and expands it even further. In this way Russia is taken as a 'new' member of the 'ring of friends'. Moreover, Russia and the EU decided to further develop their strategic partnership through the creation of four common spaces (that is, common economic space, common space of freedom, security and justice, common space of cooperation in the field of external security and common space of research education and culture) within the framework of the existing PCA (Commission of the European Communities 2004a, p. 4, 6; Commission of the European Communities 2004d, p. 1).

Though the EU-Russian border has gone through broad changes during the enlargement process, it should not be regarded simply as an extension of the existing one. First, on the EU side, the border is still in the making. The eastern enlargement of 2004 was not a linear process but rather incremental, and in many respects still awaits confirmation. Second, according to the Commission's plan the design of various EU policy instruments of cross-border cooperation and assistance will be reshaped and a New Neighbourhood and Partnership Instrument put in place after 2007. Moreover, the latest enlargement round may have caused a shift in the balance of interests in the EU, affecting its external behaviour (Kononenko 2004, p. 3). The new eastern border can be expected to remain stable. This does not, however, mean that the new border will create an impermeable 'iron curtain'. In contrast, at least from the EU's standpoint, regional cooperation between the EU and Russia has special importance in the interests of both parties (Commission of the European Communities 2004a, p. 20).

Existing Financial Instruments

The EU has developed several funding instruments to develop and support regional-level CBC in its border regions. The INTERREG Community Initiative was adopted in 1990 and was intended to prepare border areas for a Community without internal frontiers. The emphasis has changed over the years and nowadays INTERREG is the most important financial instrument on the external borders of the EU. The first INTERREG period lasted from 1990 until the beginning of 1994 and the launching of the INTERREG II period which ended in 1999. The ongoing INTERREG III period lasts from 2000 to the end of 2006. It consists of three strands; a) cross-border cooperation between adjacent regions; b) a framework of transnational cooperation, and c) interregional cooperation.[2]

The TACIS programme was launched in 1991 to provide grant-financed technical assistance to 13 countries in eastern Europe and central Asia. The current regulation is based on an understanding that cooperation is a reciprocal process encouraging a move from 'demand-driven' to 'dialogue-driven' programming. More flexibility in the TACIS structure allows potential technical assistance to be mobilized and implemented according to the capacity of each partner country.[3]

To help candidate countries duly prepare for membership, the EU launched three pre-accession instruments: PHARE, Instrument for Structural Policies for Pre-Accession (ISPA) and Special Accession Programme for Agriculture and Rural Development (SAPARD). The PHARE programme was originally created in 1989 to assist Poland and Hungary. Later it encompassed the ten candidate countries of central and eastern Europe – Bulgaria, the Czech Republic, Estonia, Hungary, Latvia, Lithuania, Poland, Slovakia, Slovenia and Romania – helping them through a period of massive economic restructuring and political change. ISPA was designed to address environmental and transport infrastructure priorities identified in the accession partnerships with the ten applicant countries of central and eastern Europe. It was established in June 1999 on the basis of a Commission proposal in Agenda 2000 to enhance economic and social cohesion in the applicant countries of central and eastern Europe for the period 2000–2006. The third pre-accession instrument, SAPARD, aims to help the ten beneficiary countries deal with problems of structural adjustment in their agricultural sectors and rural areas, as well as in the implementation of the *acquis communautaire* concerning the Common Agricultural Policy (CAP) and related legislation.

The MEDA programme is the principal financial instrument of the European Union for the implementation of the Euro-Mediterranean Partnership. It offers technical and financial support measures to accompany the reform of economic and social structures in the Mediterranean partners and is implemented by the

---

2   See <europa.eu.int/comm/regional_policy/interreg3/abc/abc_en.htm> [last access on 6 October 2005].

3   See <europa.eu.int/comm/external_relations/ceeca/tacis/> [last access on 6 October 2005].

Directorate-General EuropAid. The legal basis of the MEDA Programme is the 1996 MEDA Regulation (Council of the European Union 1996). Community Assistance for Reconstruction, Development and Stabilisation (CARDS) is a programme for the countries in the western Balkan; the programme's wider objective is to support the participation of Albania, Bosnia and Herzegovina, Croatia, Serbia, Montenegro and the former Yugoslav Republic of Macedonia in the Stabilisation and Association Process (SAP). It seeks to promote stability within the region while also facilitating closer association with the European Union.[4]

Towards a New Neighbourhood

The basic principles of EU policies for cross-border cooperation were introduced in the document Wider Europe — Neighbourhood: A New Framework for Relations with our Eastern and Southern Neighbours (Commission of the European Communities 2003a). ENP includes those neighbouring countries which do not currently have a perspective of EU membership.[5] The document also defines the main goals of its New Neighbourhood policy. First, the Union should avoid drawing new dividing lines in Europe and rather promote stability and prosperity within and beyond its new borders. Second, the EU should aim to develop a zone of prosperity and a friendly neighbourhood – a 'ring of friends' – with whom the EU enjoys close, peaceful and cooperative relations. In addition, Russia, the western countries of the Newly Independent States (NIS) and the southern Mediterranean should be offered the prospect of a stake in the EU's internal market and further integration and liberalization to promote the free movement of persons, goods, services and capital ('four freedoms') (ibid., p. 4, 12).

The most important tool for accomplishing the goals set in the Wider Europe document is the European Neighbourhood and Partnership Instrument (ENPI).[6] It will replace existing geographical and thematic programmes covering the countries concerned and places special emphasis on cross-border cooperation (Commission of the European Communities 2004b, p. 2). The development of the ENPI is a two-phased process. During the first period (2004–2006) the key objective is to build on existing progress made in coordinating the various instruments while fulfilling existing commitments and obligations regarding the current programming period up to the end of 2006 (Commission of the European Communities 2003b, p. 8). The first (transitional) period focuses on improving cooperation between various financing instruments concerned with the existing legislative and financial framework on the

---

4 See <europa.eu.int/comm/enlargement/financial_assistance.htm> [last access on 6 October 2005].

5 Algeria, Armenia, Azerbaijan, Belarus, Egypt, Georgia, Israel, Jordan, Lebanon, Libya, Moldova, Morocco, the Palestinian Authority, the Russian Federation, Syria, Tunisia, Ukraine.

6 It was at first called European Neighbourhood (Commission of the European Communities 2004a).

external borders of the EU. From the beginning of the 2007 ENPI will replace the current TACIS and MEDA programmes in the ENP partner countries.

In 2004 the European Commission took a new step to develop its neighbourhood policy by introducing the European Neighbourhood Strategy Paper (Commission of the European Communities 2004a). This document sets out the principles, geographical scope, methodology for implementation of the ENP and issues related to regional cooperation and gives guidelines concerning financial support for the implementation of the ENP. In September the Commission introduced new action plans to establish jointly defined key priorities in selected areas (ibid., p. 9–10). The EU is also planning to integrate Armenia, Georgia, Azerbaijan, Egypt and Lebanon into the NNP. In addition, the EU leaves the door open for further (geographic) expansion of the NNP by declaring that '[t]he European Neighbourhood Policy is open to Belarus and all Mediterranean countries including Libya, once they have met the necessary conditions for inclusion' (Commission of the European Communities 2004f, p. 4–5).

Despite its good intentions, the New Neighbourhood Policy includes some problematic features. First, it stresses common European values and cultural ties strengthened by a long history of living together in the border regions (Commission of the European Communities 2003b, p. 4). In some cases emphasizing common European values and identification is, however, more or less forged Europe-making from above as there are border regions where ties are not that strong and history is full of tensions and prejudices.

The second major problem is connected to the first. The creation of a 'ring of friends' leaves some countries without the prospect of EU membership.

> The aim of the new Neighbourhood Policy is, therefore, to provide a framework for the development of a new relationship which would not, in the medium-term, include a perspective of membership or a role in the Union's institutions. A response to the practical issues posed by proximity and neighbourhood should be seen as separate from the question of EU accession. (Commission of the European Communities 2003a, p. 5)

Since it is not explicitly mentioned who will eventually be left out and who might possibly be accepted as a future member, we must read between the lines. Some academics thus argue that referring to Russia, Belarus, Ukraine and Moldova as 'good neighbours' implies that they will nevertheless remain outsiders (Kramsch et al. 2004, p. 8). This problem has also been noted by the European Parliament, which in its report on the Wider Europe document points out that

> [...] we need to ensure that this 'ring of friends' approach does not turn into a crude argument between those who say that 'neighbouring country' means 'a country destined never to join the EU' and those who believe that the prospect of joining the EU is the only way of achieving closer cooperation. (European Parliament 2003, p. 17)

Since the situation on the southern external border is apparently clear, speculation about possible future members focuses mainly on the eastern border of the Union.

Accession has been ruled out, for example, for the non-European Mediterranean partners (Commission of the European Communities 2003a, p. 5). Even though this definition is problematic, for example, in the case of Turkey,[7] the European Union will, thus, in the future be bordered in the south by the Mediterranean; no final decision has been taken about the eastern border. It is therefore possible that in the future two sets of 'rings of friends' will exist on the external borders of the European Union: a ring containing those who will later be accepted into the enlarging Union and a second composed of those who will be left outside this enlargement ring. This division would put different border regions in different categories and could also reduce motivation for CBC in regions belonging to the 'outside' category.

Thirdly, the idea that a joined and unified effort is needed, since the member countries are not well-equipped to address the new challenges of CBC, can be understood as part of policies targeted at cutting the authority of the member states in defining priorities and allocating funds. In spite of the promises of empowering border regions with new instruments, the goal of streamlining the decision-making powers that today range from the local and regional to the national and EU levels can even be considered as a threat to today's dispersed but independent regional institutional structures.

Regional Instruments of CBC – from Internal Boundaries to External Borders

The latest round of enlargement has profoundly changed the nature of CBC on the external borders of the EU. Prior to the enlargement round, the most intensive cooperation across the eastern border was part of the pre-integration of future member countries. The EU wanted to prepare and integrate accession countries before signing the membership agreements. The pre-integration phase was supported through numerous financial instruments dedicated to border regions (especially PHARE CBC but also through INTERREG and TACIS). Following enlargement the pre-integration aim no longer exists on most of the eastern external borders. Instead of lowering the fences, the goal is rather to establish new cooperation forms across a (more) stable eastern border which with time would meet the Schengen regulations. In this situation CBC between EU member countries and Russia has gained special importance. For the EU, greater regional cooperation in eastern Europe will bring substantial benefits, and in the EU documents the participation of the Russian Federation as a partner in regional cooperation has been willingly encouraged (Commission of the European Communities 2004a, p. 20).

This development has also set the elaboration of regional institutional instruments of CBC in a new light. Regional institutional forms for promoting CBC have been actively constructed since the second world war, and since the late 1950s

---

7   'In the Mediterranean region, the ENP applies to all the non-EU participants in the Euro-Mediterranean Partnership (the Barcelona process) with the exception of Turkey, which is pursuing its relations with the EU in a pre-accession framework.' (Commission of the European Communities 2004a, p. 7).

Euroregions have become the most noticeable model of this work.[8] Perkmann (2002, p. 104) defines Euroregions as a more or less institutionalized collaboration between contiguous subnational authorities across national borders. Perkmann's classification is broad and could include other (institutionalized) forms of CBC as well.

Euroregions were originally founded on the internal borders of the European Community, but in the 1990s they rapidly started to materialize in central and eastern Europe, and especially across the eastern external border of the EU. There were two main reasons for this trend. First, the collapse of the iron curtain opened the eastern external border for cooperation between formerly closed entities. The second factor was the EU pre-integration policy, which was soon established to prepare possible new member countries for accession. The need to promote development and cooperation across the new eastern border created a need for stable institutional structures. While the local and regional administrative structures, especially in the countries that had lived under communist rule, were often extremely weak, Euroregions sought to answer this challenge and give regional-level CBC a more authorized nature (Virtanen 2004, p. 126.)

In terms of activities, Euregions have been mainly concerned with administrative matters that demand cross-border coordination at the local level. Traditionally such coordination concerned issues involving spatial planning, transport and environmental externalities (Perkmann 2002, p. 105). The main goal of these organizations was to promote cooperative initiatives that address specific economic, environmental, social and institutional problems affecting their respective region (Scott 1999, p. 9). Problems related to this type of questions were soon actualized after the collapse of the iron curtain but in many cases the social and institutional problems rather hindered than encouraged cooperation across the border.

Though circumstances varied considerably in different border regions, it was possible to find similar obstacles deriving from weaknesses in institutional structures and economic difficulties along the EU's entire eastern border. Especially on the Finnish-Russian border, the historical burden and weaknesses in institutional structures and the poor economic situation on the Russian side limited the creation and execution of cooperation projects (Shlyamin 2004, p. 130). The role of the Euroregions was thus shaped not only by the will to promote mutual interests but also to reduce elements that limit the possibilities for successful cooperation. To facilitate CBC and better coordinate its financing on the Finnish-Russian border, Euregio Karelia was founded in 2000. At that point it was the first Euroregion to be established on the land border between the Russian Federation and the EU.

Euregio Karelia as a Tool of European Border Policies

Euregio Karelia covers 700 km of land border between the EU and Russia and has a population of 1.4 million people. The overall area of the Euregio is 263,667 km²,

---

8   The first Euregio was founded in 1958 on the border between Germany and the Netherlands.

180,500 km² belonging to the Karelian Republic and 83,000 km² to Finland, and it consists of four regions: the provinces of North Karelia, Kainuu and North Ostrobothnia on the Finnish side and the Republic of Karelia on the Russian (Euregio Karelia 2000)

Euregio Karelia is described as a continuous process in which cooperation seeking a common goal takes place at a concrete level on both sides of the border. Its objective is to facilitate interaction across the border, to increase welfare on both sides as well as to promote democracy. The framework of Euregio Karelia covers cross-border cooperation in the fields of business, environment, tourism and culture (Euregio Karelia 2000).

Documents published by Euregio Karelia emphasize three similarities between the Finnish and Russian partner regions: a common border, a common nature and common cultural and cooperation traditions, as well as experiences gained on both sides of the border in EU programmes and the coordination of different programmes. The first of these similarities – the border – is in principle simple and clear. It can be seen as a dividing line or as zone of cooperation, depending on one's point of view, but still forming a common border. The second similarity, the common nature, too,

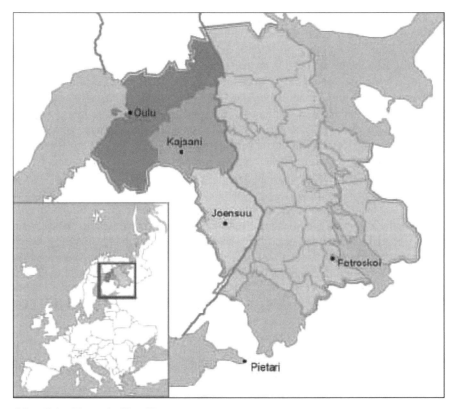

Map 8.1   Euregio Karelia

is straightforward and leaves little room for debate. It is based on a geographical location ensuring a similar climate and similar natural features and landscapes, mainly forests and lakes.

The third similarity, the idea of common cultural and cooperation traditions, is more open to interpretation because of the complex history of the Finnish-Russian border. Over the past centuries the borderline has changed many times. After the second world war the Finnish-Russian border was practically closed and remained strictly guarded until the collapse of the Soviet Union. During the cold war the framework of cross-border cooperation was created at the state level, and Finnish-Russian cooperation was guided by the Agreement of Friendship, Cooperation and Mutual Assistance concluded in 1948. This was the basic document of postwar foreign policy in Finland, and it controlled Finland's relationship not only with the Soviet Union but also with western countries. After the war parts of Finnish territory were ceded to the Soviet Union and some 400,000 people had to be evacuated and resettled in other parts of Finland. As the ceded part of Karelia was emptied of the Finnish population, people from various parts of the Soviet Union moved into the area during the 1940s and 1950s, and for the most part place names and other symbols were russified (Forsberg 1995, p. 207–9).

Despite the burdens in cultural and cooperation traditions, visions connected to Euregio Karelia have been positive. The leading figures behind the initiative, Tarja Cronberg and Valeri Shlyamin (1999), had already recognized during the initial stage that the benefits would also involve changing attitudes towards the border and that the cooperative zone would reduce the historical burdens or at least provide an alternative mental frame for collaboration.[9] The future of the Finnish-Russian border would no longer be that of a dividing line but, instead, Euregio Karelia was to show that borders are also bonds between people, communities and regions.

Since Euregio Karelia was the first Euroregion on the land border between the EU and the Russian Federation, the key figures behind the venture willingly promoted it from the beginning as a European model (Liikanen 2004). It was seen as a pilot project for future joint administrative structures between the EU and Russian regional authorities. The idea was that the structures developed in this region would with time gain broader European significance. From the Finnish perspective, the institutional forms adopted on the Russian border were seen as exporting 'border know-how'. They would generate a model or at least a set of experiences that could be useful for the elaboration of European border policies after the eastern enlargement (Cronberg 2000, p. 170–83).

The role and importance of the Euroregions were recognized not only in the Euregio Karelia region but also by the European Commission.

Euroregions involve concrete cooperation between regional and local authorities on both sides of the border, which can in time lead to substantial and effective links across the

9   Tarja Cronberg and Valeri Shlyamin were the first heads of the Management Committee of Euregio Karelia.

borders. They can promote common interests and thus strengthen civil society and local democracy as well as having beneficial effects on the local economy. (Commission of the European Communities 2004a, p. 21)

According to the Commission the development of integration and cohesion within the enlarged European Union and its border regions is an immense challenge and requires considerable action. Euroregions have their role in this process as mediators and coordinators. A combination of different funding opportunities and the building of an administrative governance structure to help project-based cross-border cooperation is also one of the main aims of Euregio Karelia (2000).

In future the introduction of the New Neighbourhood policies will directly affect both the regional and the European dimension of Euregio Karelia. To some degree its regional role will probably be promoted by the adaptation of ENPI, and the new policy framework is likely to further emphasize its role as a European model. At the same time the streamlined framework can limit the independence of the local actors and even weaken the legitimacy of its activities if it ignores the horizon of the local actors. In this sense the roles and influences of the Euroregions on the external border of the European Union are not limited to the local scale but their future is bound to larger processes of policy reformulation within the EU and on its (eastern) external borders (Virtanen 2004, p. 133).

While terms like cooperation and integration are commonly invoked in a positive manner, practice shows that successful cross-border activities do not arise simply by adjusting administrative and legal regulations. Rather the inhabitants of border regions must also view their 'border opposite' in a light where cooperation, both formal and informal, is seen as advantageous and necessary (Bucken-Knapp and Schack 2001, p. 16). This is one of the key challenges to Euregio Karelia as well as to the reformulation of EU border policies.

European Model – European Identity?

The end of the cold war has changed images of Europe and its borders dramatically. The reconceptualization of borders and cross-border interaction has been particularly profound in the case of Finland and its eastern border. In terms of international politics the border has lost its meaning as a dividing line between two rival social and political systems, or as a 'Finlandized' grey zone between them. But this has not meant that the Finnish-Russian border has been stripped of supranational definitions. Since 1995, when Finland joined the EU, the border has become an object of manifold 'Europeanization', and cross-border interaction has been reconceptualized in terms of European integration and EU politics.

During the cold war the border between Finland and the Soviet Union was heavily guarded on both sides. Cooperation and trade were administered by bilateral agreements between the two states. Border crossing was subject to tight visa regulations at only a few crossing points. From the regional and local perspective the border was virtually closed. Since the collapse of the Soviet system the border has remained heavily guarded, but the forms of cross-border cooperation have changed and a new scale of interaction has emerged. Today regional and local actors are taking an active role in cross-border cooperation. Regional administrative units, economic enterprises and civil society organizations cooperate directly across the border.

Interstate relations between Finland and Russia constitute, even in the new situation, the basic framework for cross-border cooperation. However, from a European perspective the key question is no longer simply nation-level interaction, but how the new supranational and regional dynamics are mediated in the process. In this twofold transformation, that is simultaneous internationalization and regionalization, an initiative like Euregio Karelia has become part of a manifold identity politics, the construction and reconstruction of European, national and regional identities.

As stated above the key figures behind the Euregio Karelia initiative promoted the new institutional structure as a new European model from the very beginning. The idea was that, as the EU enlarged eastwards, joint administrative structures with Russian regional authorities would gain broader European significance (Cronberg and Shlyamin 1999, p. 325, 326). This argument was, however, not limited to establishing a new kind of border regime but rather it was introduced in terms of a new kind of cross-border region-building. In the planning phase of Euregio Karelia, Tarja Cronberg, head of the Regional Council of Finnish North Karelia, anticipated that '[c]ommon decision-making procedures and common funds [would] create a foundation for establishing new border region identities'.

> The Euroregions are bridges between countries. They form new links between former enemies based on culture, sometimes a common language and a common history. In a way they are crucial for developing the European community, and they help to promote integration and a common identity for the regions. (Cronberg and Shlyamin 1999, p. 325–6)

In this respect, the aims of the initiators were not limited to organizing a regional border regime, but touched on key questions of European and national identity politics. In their article 'Euroregion Karelia – A Model for Cooperation at the EU External Borders', Tarja Cronberg and Valeri Shlyamin, minister for external relations of the Karelian Republic, initially set the goals of the project in fairly concrete terms. The coordination of INTERREG and TACIS programmes at the regional level was presented as the core of the new administrative model. Yet even at this stage, easing border crossings and increasing economic, social and cultural cooperation were discussed in connection with questions of security and attitudes to the border:

> The benefits of Euregio Karelia for the EU would comprise a more intensive and effective use of funds, which now flow to both sides of the border and which are not coordinated. The benefit for Russia would be increased cooperation across the border, which later would also imply more economic activities [...] From the Finnish side, the benefits would comprise changing attitudes towards the border and removing the historical burden. The Karelian question in Finland is on the agenda and a number of people work for actual

physical changes in the border. A cooperative zone would remove the historical burden or at least provide a different prospective. (Cronberg and Shlyamin 1999, p. 28–9)

The benefits were many: for the EU, the coordination of aid programmes; for Russian Karelia, economic progress; and for Finland, stability and the removal of historical burdens. The final aim was expressed in rather grandiose terms – even for a paradiplomatic document reflecting relations between subnational governments:

> By providing a continuous process for cooperation towards more integrated structures in economic and social development, Euregio Karelia would show that borders no longer separate but rather form both historical and future-oriented bonds between people, communities and regions on both sides of the border (Cronberg and Shlyamin 1999, p. 29).

For the initiators of the Euroregion model, refashioning mental borders was obviously a major aim behind the initiative – at least on the level of declarations. In this respect the obstacles have probably proven to be larger than expected. In the case of the Karelian Republic the consolidation of the Russian nation-state has since strongly affected the political climate, and from the perspective of Russian nationalism cross-border region-building can easily be seen as a source of discord, or even a threat. This has led to the paradox that, while in Finland the concept of Euregio Karelia was promoted as an alternative to nostalgic postwar Karelianism (and to the marginal militant Karelia activism as well), in the Russian discussion it has sometimes been connected precisely to ideas of *revanche* in regard to the ceded areas.

As the results of a recent study carried out within the framework of the EXLINEA project indicate, attitudes towards cross-border cooperation in Russian border areas are for the most part very positive. Similarly, on the Finnish side, the actors involved in cross-border interaction see the future prospects very positively. This positive engagement has, however, not been followed by strong patterns of common identification, neither in the sense of common cross-border identity nor in terms of broader European identification. Instead, the common ground for mutual understanding and solidarity seems to lie closer to the level of everyday practices, in lessening the obstacles to cooperation and in the sense of the ability to affect the conditions of cross-border interaction through national and regional political channels.

Conclusions

Whether the concept of the Euroregion can become significant as a cooperation model while awaiting the adoption of the New Neighbourhood policy probably depends mainly on how well it can promote practical everyday cooperation and on the degree to which it can strengthen the feeling that local actors have a say in the administration of cross-border interaction. In this sense the desire for a more efficient policy-driven approach, which the adoption of ENPI represents, is likely to find support on the regional level, and the Euroregions can in principle become a functioning part of the new system.

On the other hand, if the new policies are promoted as a new streamlined European model with great expectations of strengthening common European identification, adopting the policy frame will probably meet with more suspicion at the regional level.

In this regard the question comes back to broader lineages of European politics. As in the case of the Constitution, the benefits of streamlined and more efficient European government will perhaps not be accepted willingly if there is a threat of weakening the possibilities of the citizenry to affect policies through national and regional channels.

In the Finnish case the idea of 'one framework, one instrument' European border policies seems to suggest that future post-enlargement border policies between the EU and Russia should not be planned primarily on the basis of earlier Finnish experiences of the border. In fact, it seems possible that new solutions on the Finnish border are more likely to have their origin in solutions to problems of cross-border interaction between Russia and the new EU member countries. If the new policies and policy instruments mean empowering regional actors and regional institutional structures as part of a new streamlined European chain of command, the role of the Euroregions could hardly have broader regional significance or strengthen the legitimacy of EU policies. These goals are more likely to be achieved if regional institutional structures gain more the role of mediators between European, national and regional aspirations and interests connected to the border, if they allowed the citizens to participate in the cooperation not just as part of a grand European vision but on the basis of their own national and regional interests.

In the case of Euregio Karelia the question is not primarily whether this form of cooperation will become part of new semi-official European structures of cross-border cooperation. The uniqueness of Euregio Karelia lies not in the fact that it represents the tendency of the border withering away in the foreseeable future. Rather its role can be found in anticipating the simultaneous tendencies of internationalization and regionalization of the Finnish-Russian border. The border will remain and there will in the future also be different supranational, national and regional interests that are associated with the Finnish-Russian border. In this situation the potential future role of the Euregio is probably not so much to promote interaction according to a predefined European policy framework, but rather to open a channel to mediate between supranational, national and regional ambitions. In regard to the New Neighbourhood Instrument the crucial question is maybe not so much whether it is possible to introduce new 'European-type' institutions on the external borders but whether the kind of institutional innovations that Euregio Karelia represents will be able to coordinate the complicated interplay between different spatial levels and mediate between European, national and regional interests and aspirations connected to the border.

PART IV
Evolving Cooperation Frameworks and Cross-Border Regional Development

Chapter 9

The Impact of EU Enlargement on Moldovan-Romanian Relations

Alla Skvortova

Moldova and Romania have a long history of common ethnic, cultural and linguistic ties. However, these commonalities have not always been helpful in overcoming political tensions. This is also true for the period after 1991, when Moldova regained its independence. Romania's invitation to join the EU accession process has helped Moldovan decision-makers realize that their country might benefit from Romanian support and expertise in their own negotiations with the EU. Within this context, the European Commission's Strategy on Wider Europe – Neighbourhood has had a very positive impact; this policy strategy emphasizes much more active cooperation between EU candidate countries and their neighbours in order to promote security and stability at the EU's future borders. It was the European Commission that initiated the development of the Neighbourhood Programme with Romania and Moldova and, in this way, put new life into the relations between the two countries. This essay will discuss the possible impacts of EU enlargement and EU policies on the further development of the relations between Moldova and Romania. In doing this, the author will provide a short overview of the history of these relations and will, furthermore, trace their development since Moldova's independence.

Political Contexts For Cross-Border Cooperation

Relations between Romania and Moldova during the last fifteen years have followed a rather confused trajectory. The Romanian revolution of December 1989 as well as political and ideological changes produced by Gorbachev's perestroika allowed for open borders between the two countries as well as free travel and contacts. There was also open recognition that the Moldovan language was not much different from Romanian, while some Moldovans declared that they were in fact ethnic Romanians.[1] Many thought that unification of the two countries was inevitable. In 1990 the strongest, most numerous and most powerful political force in Moldova at the time – the Popular Front – included the objective of unification in its programme. The arguments in favour of unification were drawn from a rather chequered history.

---

1   On the discourse on the Moldovan-Romanian identity issue see King (2000).

Similar languages, traditions, folk costumes, music and dances, rural architecture, common heroes and writers – all these result from the fact that the medieval Moldovan principality incorporated territories on both the left and right bank of the Prut River (1389–1812). After the Russian-Turkish War of 1806–1812 the eastern part of the Moldovan principality, situated between the Prut and Nistru Rivers, was annexed by the Russian Empire as a war trophy, while the rest remained under the Ottoman Empire, with considerable autonomy guaranteed by the Russian Empire. In 1862 the Moldovan and Wallachian principalities merged into what came to be called the Romanian principality, later renamed into the Kingdom of Romania. After the Russian-Turkish war of 1877–1878, in which Romania actively participated, the kingdom obtained its independence from the Ottomans. To summarize, since the division of the Moldovan principality in 1812, the territory of modern Moldova has successively belonged to the Russian Empire (1812–1917), Romania (1918–40 and 1941–44) and the USSR (1940–41 and 1944–91) before becoming the present Republic of Moldova, while Moldova beyond the Prut has always been part of Romania since 1862 (Skvortova 2002). Supporters of unification described the years when Moldova belonged to Romania as the most glorious in the history of the country, whereas opponents, identified as ethnic Moldavians, feared they would become a national minority in the case of unification.

At the time there were signs that some political groups in Romania would welcome unification too, though officially the issue was never raised. Romania was the first country to recognize the new Republic of Moldova, only a few hours after its declaration of independence on 27 August 1991. Moldova was mentioned in all Romania's strategic programmes and was supported when it came to establishing diplomatic relations with other countries. Moldova also received military and cultural assistance, and thousands of Moldovan students were offered scholarships at Romanian universities (Serebrian 2004, p. 149). At the same time, the Romanian president Ion Iliescu reintroduced into the political discourse the notion of 'two Romanian states' that were to build 'privileged' relations of friendship. In Moldova, this was interpreted as a long-term commitment to unification.

The parliamentary elections held in Moldova in February 1994 brought a majority for the Agrarian-Democratic party whose declared aim was to strengthen the independent state and Moldovan identity as well as to promote Moldovan as a language distinct from Romanian (Crowther 1998, p. 150–51). The new government opposed the concept of 'two Romanian states', let diplomatic relations with its neighbour cool and significantly reduced contacts. After the elections of 1998 political forces ranging from the centre to the right had to form a single bloc in parliament to outweigh the Communist faction. Since the right, and to an extent the centre-right, were pro-Romanian, relations between the two countries improved and economic, political and cultural contacts intensified, leading to the signing of dozens of agreements, treaties and conventions in different areas, the implementation of 1,096 joint ventures by 2000, the disbursement of special Romanian government funds to Moldova and the offer of significant Romanian support for cultural and educational institutions (Serebrian 2004, pp. 142).

However, during the same period, the EU enlargement policy had a controversial influence on relations between Chisinau and Bucharest. In the aftermath of the Declaration of the Helsinki Summit (December 1999), for instance, the pro-Romanian political parties made an attempt to revive the idea of unification by stressing the prospect of eventual integration into the European Union, whereas the Moldovan government felt offended by the fact that only Romania had been invited to join the accession process (Skvortova 2001, p. 106). Soon afterwards the Moldovan leadership was, however, having second thoughts and declared that all necessary steps would be taken to benefit from Romania's new status as an EU accession state. Official rhetoric reiterated that the country should find new and more efficient forms of collaboration with Romania and try to take advantage of the latter's new position.[2] Invitation of Romania to the accession process was seen as an advantage because it offered, above all, a possibility to learn from the Romanian experience about the reforms required by the EU. In an instance of wishful thinking, Romania's invitation was even interpreted as a sign that Moldova would follow suit, as both countries shared similar political, economical and social problems.[3] The political forces favouring unification tried to take advantage of this situation. They were well aware that the overwhelming majority of Moldovans were not willing to renounce independence and therefore argued that unification with Romania would offer Moldovans citizenship of an EU country.[4]

As for Romania, its interest in Moldava almost disappeared with its new status. The country's very limited resources were completely absorbed by the accession process. But it still had to comply with the EU requirement that candidates settle their problems with neighbouring countries. Romania thus had to conclude basic treaties with its neighbours, including Moldova. To facilitate the ratification of a treaty with Moldova, Romania in 2000 committed $ 1,2 million to 'support the project of strengthening the privileged partnership between the two countries' and to 'continue political, economic and diplomatic support of the Euro-Atlantic orientation of Chisinau'.[5] The basic treaty was signed in May 2000, during an official visit of the Romanian minister of foreign affairs, Petru Roman, to Chisinau. The text of the treaty has not been made public but received negative comments from pro-Romanian political parties because it did not reflect Romania's special interests in Moldova. These parties expected that the idea of the existence of two Romanian states as well as a statement on the Molotov-Ribbentrop Pact would be included in the text – as Romania had long insisted. This request had been one of the main obstacles to an earlier signature. But once EU accession had been offered to Romania, Bucharest changed its approach to the 'Bessarabian question': the desire and readiness, though never officially recognized, to accept this territory as an historical part of Romania

---

2 *FLUX*, 30 and 31 March 2000.
3 Interview with Emil Ciobu, Moldovan ambassador to Romania, *FLUX*, 30 March 2000.
4 *Nezavisimaia Moldova*, 24 March 2000.
5 *Nezavisimaia Moldova*, 26 March 2000.

was replaced by an attitude similar to that adopted towards other neighbouring states. In an interview given at Chisinau, the Romanian minister of foreign affairs stressed that the finalisation of the treaty had been stimulated by the opening of accession negotiations with the EU.[6]

Political forces in Moldova advocating reunification with Romania thus lost an important component of their ideology. From the EU's point of view this could be interpreted as a positive move strengthening the region's security situation, since opponents of reunification now formed a majority.

During early parliamentary elections in February 2001, the Moldovan Communist Party won 71 out of 101 seats and thus obtained the right to elect the president and appoint the government. Contrary to widespread expectations, the new leaders declared European orientation to be the priority of Moldovan foreign policy, actively worked for Moldova to be accepted into different European organizations and even expressed hopes that Moldova would be part of the accession process. To general surprise, Vladimir Voronin's pro-European rhetoric was much more precise than his predecessors'. The new government suggested to start negotiations on a free trade agreement with the EU, the establishment of an EU representation in Chisinau and Moldova obtaining the status of an associate member of the EU.[7] A draft foreign policy concept that was to replace an earlier one adopted in 1995 was published in April 2002. It contained a chapter on 'European integration as the prior strategy goal'. This was the first ever Moldovan official political document that described in detail reasons why Moldova needed to join the EU and the actions necessary to achieve this goal.

These aspirations for European integration could not but influence Moldova's policy towards Romania. The new government thus chose a more pragmatic way of balancing between East and West and demonstrated its willingness to continue the country's 'special relations' with Romania. The Moldovan ministry of foreign affairs stated that relations with Romania would continue to represent 'one of the prior objectives of Moldovan foreign policy'.[8] The official visit of the Moldovan president to Romania was punctuated by statements about this special relationship as well as closer cooperation. Iliescu and Voronin agreed that relations between the two countries should be based on 'pragmatism'.[9] A special Commission on Relations with Romania was created, and Romania was one of the main supporters of Moldovan membership in the Stability Pact.

However, despite this, Moldovan-Romanian relations deteriorated. The Romanian president Iliescu sustained the idea that Moldova was 'the second Romanian State' and suggested to introduce a reference to the Molotov-Ribbentrop pact into the basic treaty concluded between the two countries, a demand which was unacceptable for

---

6   *FLUX*, 3 May 2000.

7   Popescu, Nicu, 'Euromanie comunista? ', *Jurnal de Chisinău*, 28 September 2001.

8   *Moldova Suverana*, 24 March 2001.

9   Neukirch, Claus, 'Conceptul politicii externe a R.Moldova este ambiguu si dificil de urmat', *FLUX*, 21 December 2001.

Chisinau as it appeared to threaten Moldova's sovereignty.[10] The Romanian prime minister, Adrian Nastase, went even further by denying the very need of a basic treaty with Moldova. He stated that Romania 'recognized Moldovan independence due to some considerations',[11] adding that 'the treaty should provide for the possibility for the population of both countries at an opportune moment to make a decision distinct from the one made initially', that is unification.[12]

A further worsening of the relations between the two countries was due to the conflict between the Moldovan and Bessarabian Metropolitans, which had reached the European Court for Human Rights. The Romanian prime minister Adrian Nastase cancelled a planned official visit to Moldova after the Moldovan minister of justice, Ion Morei, accused Romania of expansionism and interference into the internal affairs of Moldova by supporting the Metropolitan of Bessarabia in a speech addressed to the European Court.[13] Some observers advanced that the Romanian government ended the special procedure for granting Romanian citizenship to Moldovans not because of EU requirements but as a response to the Moldovan government's position on the Metropolitan issue and its attempt to introduce the Russian language into the curriculum of Moldovan elementary schools.[14] The Romanian prime minister, Adrian Nastase, compared the latter step with Chisinau's continuous attempts at an 'elimination of Romanian identity' in the Republic of Moldova. Nastase supported protests against the introduction of the Russian language in primary schools and against the replacement of an 'History of Romanians' by an 'History of Moldova' in the overall education system.[15] This time Moldova aspired to get support from the European Union by using in its rhetoric the language of the European Union. The prime minister, Vasile Tarlev, stated that Nastase's position 'prove[d] the disrespect by the Romanian administration of European norms and standards, democratic principles and, especially, of principles of international law regarding relations between states, rights of each country to political or other kind of self-determination without interference by other states'.[16]

Indeed, the accession status of Romania helped resolve the conflict. President Ion Iliescu advised those in Romania who supported the protesters in Moldova that their activity could affect the image of Romania and interfere with its integration into NATO and the EU.[17] As a result the Romanian parliament, fearing repercussions on the negotiations for NATO membership, voted against the 'Bessarabia' resolution proposed by the Party of Greater Romania who wanted increased financial support

---

10 Ibid.

11 *FLUX*, 20 September 2001.

12 Rotaru, Cristina, 'Tratatul moldo-roman nu va fi ratificat de Parlamentul Romaniei', *FLUX,* 20 September 2001.

13 *FLUX,* 4 October 2001.

14 Tkachuk, Timur, 'Utrachennye illiuzii: teperi i vo vneshnei politike', *Moldavskie vedomost,i,* 24 October 2001.

15 *Nezavisimaia Moldova,* 18 January 2002.

16 *Moldavskie vedomost,i,* 30 January 2002.

17 Pozitia lui Ion Iliescu, *Tara* 2 February 2002.

for 'patriotic, Romanian and unionist' publications in Moldova and the Bukovina.[18] 'Pragmatism' became the key word of Romanian foreign policy, and it was decided to focus on economic relations and continue to support Moldova's efforts towards EU integration, once senior European officials expressed their concerns about the political situation in Moldova and its tense relations with its neighbours (Serebrian 2004, p. 151).

Economic relations

Changes in the political relations between Moldova and Romania have affected the economic relations between the two countries. Despite a comprehensive legal framework for economic cooperation,[19] Romania has reacted to the political conflict by erecting sanitary, veterinarian, technical and other barriers for Moldovan goods each time political tensions between the two countries rise. From time to time Romania thus prohibits imports of Moldovan meat, dairy, eggs, and tobacco on the grounds that these commodities do not meet EU standards, although neither country is an EU member. At the same time Romania has created obstacles for the import of Moldovan steel, concrete, sugar, wine and other goods. In violation of an earlier agreement Romania has excised duties and other taxes on Moldovan alcohol and tobacco products many times higher than for similar local products.[20] This evidently is not in line with EU recommendations on the intensification of cross-border cooperation.

---

18 'Parlament Rumynii ne poddalsia emotsiam', *Nezavisimaia Moldova,* 19 February 2002.

19 The legal basis for regulating flows of goods and people at the border between the Republic of Moldova and Romania includes the following documents:

- The Free Trade Agreement between the Government of the Republic of Moldova and the Government of Romania (17 November 1994);
- The Agreement between the Government of the Republic of Moldova and the Government of Romania on Customs Cooperation and Mutual Administrative Assistance for the Prevention, Investigation and Repression of Customs offences (it sets forth simple border crossing procedures for goods and people by organizing joint control, efficient information exchange, etc.; 15 October 2000);
- The Agreement between the Government of the Republic of Moldova and the Government of Romania on the Readmission of Foreigners (27 July 2001);
- The Agreement between the Government of the Republic of Moldova and the Government of Romania on Mutual Travels of Citizens (11 September 2001);
- The Protocol between the Frontier Guard Department of the Republic of Moldova and the Frontier Police Inspectorate of the Ministry of Home Affairs of Romania on Mutual Travels of Citizens (27 September 2001)

By 2003 the bilateral legal framework ensuring the free movement of goods and labour consisted of 50 agreements at the national, governmental and ministerial levels.

20 'Moldova za svobodnuiu torgovliu', *Ekonomicheskoie Obozrenie – Logos Press,* 26 October 2001.

Table 9.1   Trade between Romania and Moldova in 1998–2004 (in million US dollars)[1]

	1998	1999	2000	2001	2002	2003	2004
Total	173,688	122,701	157,317	131,537	147,270	188,416	263,400
Exports from Moldova	60,810	41,187	37,821	37,941	56,712	90,230	98,900
Share in total exports (%)	9.6	8.9	8.0	6.7	8.8	11.4	—
Imports to Moldova	112,878	81,514	119,496	93,596	90,558	98,186	164,500
Share in total imports (%)	11.0	13.9	15.4	10.5	8.7	7.0	—
Balance for Moldova (%)	–47.1	–49.5	–68.3	–59.5	–37.4	–8.1	–39.9

Source: Statistical Yearbook of the Republic of Moldova (Chisinau: Statistica, 2004, pp. 589, 591 and 595) for the years 1998–2003; Customs Department of the Republic of Moldova (*Ekonomicheskoie Obozrenie – Logos Press*, 18 February 2005) for 2004.

1   Data on 1998–2003: Statistical Yearbook of the Republic of Moldova, 2004. Chisinau: Statistica, 2004, pp. 589, 591, 595; data on 2004 provided by the Customs Department of the Republic of Moldova: *EO - LP*, 18 February 2005.

Moldova's main exports to Romania are foodstuffs, alcohol beverages, textiles, clothing, machines and equipment, electric equipment and construction materials.[21] In 2002 Moldova ranked twentieth on a list of Romanian economic partners. The volume of Romanian exports to Moldova in 2002 made it the third-biggest importer, after Russia and the Ukraine, and in 2003 the second-biggest after Russia.[22] The trade balance between the two countries is negative for Moldova.

During a visit by the Romanian foreign minister, Mircea Geoana, to Chisinau in April 2003 both parties made statements in favour of boosting Moldovan-Romanian relations and promised the signing of a basic treaty by the end of the year. One of the direct results of this visit was the meeting of interministerial committees on bilateral cooperation in which representatives of the ministries of Economy and of

21   *Moldova Suverana*, 1 April 2004.

22   *FLUX*, 23 April 2003; however, according to Romanian official data, Romania ranked only fifth, after Russia, Ukraine, Italy and Germany(see <mae.ro/index.php?unde=doc&id=5 664&idlnh=1&cat=3>).

the National Banks from both countries participated. The meeting approved an action programme in the field of economic cooperation between the two countries.[23]

The significant improvement of relations at the political level since the Romanian presidential elections of 2004 might thus facilitate the future development of economic relations between the two countries. However, Romania's accession to the EU will inevitably and significantly reduce the volume of imports from Moldova, unless the latter will manage to implement EU requirements on certification and standards.

Borders and Cross-Border Travel

The border between Moldova and Romania extends for 683 km along the Pruth River. When the (then Soviet) border was opened after the Romanian revolution, huge numbers of people crossed the border in both directions, but particularly from Moldova into Romania. Moldovans visited relatives and historic places, engaged in petty trade and, most significantly, put to a test their 'kinship' with Romania. Gradually these cross-border flows decreased, and the number of Moldovan visitors to Romania was reduced by 50 per cent. Differences with Romanians, their language and culture turned out to be substantial and led to disillusionment. In the meantime petty trade was no longer profitable, and the political changes in Moldova did not encourage active contacts. Romania, on the other hand, did not introduce a visa regime for Moldovan citizens as did other EU candidate countries in 2000–2002, though the transparent border with Moldova was of serious concern to the Romanian government, especially as Romania's acceptance into the Schengen agreement was delayed.[24] However, since 1 July 2001, Moldovans require a passport to enter Romania and travel has become more difficult, even though Romania allocated $ 1 million to partially cover the cost of a passport for certain categories of Moldovan citizens.[25] Since then Romania has announced that it would have to introduce a visa regime for Moldovans in 2007, before becoming a member of the European Union. The number of border crossings between Romania and Moldova currently follows the same trend as that observed immediately after the opening of the border: visits to Romania exceed by many times those in the opposite direction, though the overall volume of travellers has significantly decreased.[26]

---

23 *Flux*, 23 April 2003.

24 Budescu, Carolina and Ion Iliescu: 'R.Moldova are probleme la frontiera de Est', *FLUX,* 31 August 2001.

25 *Kishinevskie Novosti*, 3 August 2001.

26 While in 1996 39,004 persons from Moldova traveled to Romania with the assistance of tourist agencies, only 4,347 Romanians did the same. In 2003 the figures were: 9,105 from Moldova to Romania, and 2,381 from Romania to Moldova. It should, however, be stressed that official statistics only take account of travel organized through tourism agencies, and no data exist on the much higher numbers of Moldovans who otherwise enter Romania to

Throughout this period the EU offered significant support to Moldova to facilitate cooperation with Romania and increased cross-border flows of goods and persons. Through the TACIS programme a number of projects such as building, equipping and refurbishing the customs and border posts in Ungheni, Leuseni and Giurgiulesti were implemented in order to bring them into line with EU standards. Originally, projects such as 'Enhancing border management in Ukraine, Moldova and Belarus', 'Fight against drug trafficking and drug abuse in Belarus, Ukraine and Moldova', 'Strengthening the Asylum System in Ukraine and Moldova' and 'Support to the Moldovan Border Guards Department and to border posts' were aimed at strengthening the future EU border, but at the same time they created better conditions for cross-border trade and travel.

One of the issues that raised serious EU concerns in early 2000 was the massive acquisition of Romanian citizenship by Moldovans. Romanian law – along the lines of the Baltic legal argument on state continuity – allowed for dual citizenship, including for those Moldovans whose ancestors were Romanian citizens before 1940. Romanian citizenship offered several economic advantages: as Romanian citizens Moldovans had the right to bring into the country certain goods without paying customs duties. But it was above all free travel and the right to work within the Schengen territory once Romania would join the EU that boosted the number of those willing to apply for Romanian citizenship.[27] While the Romanian authorities have never disclosed data on the number of Moldovans holding a Romanian passport, Moldovan sources from the Department of Statistics and the State Migration Service estimate their numbers at some 300,000 to 500,000 (Gheorghiu and Gutu 2004, p. 205). Analysts, politicians and decision-makers have suggested several scenarios for the time when Romania will have become a EU member; in their opinion, this situation could represent a serious threat to EU security and stability. In late 2001, probably under the pressure of the accession process and EU concerns, Romania put an end to the procedure and then, in 2002, made it much more difficult and costly, although this has not discouraged Moldovans from applying for citizenship.

Regional and Cross-Border Cooperation Patterns

Both Moldova and Romania are members of a few international and regional organizations, such as the World Trade Organization (WTO), the Stability Pact for South-Eastern Europe (SP), the South-Eastern Cooperation Initiative (SECI) and the Black Sea Economic Cooperation (BSEC), though their role in these organizations and the benefits they receive from participation are not the same. In the current context good neighbourhood relations with Romania are one of the main conditions for obtaining increased benefits from cooperation within these regional organizations.

---

sell agricultural produce, to shop, to study, to visit historical sites or to spend their vacations (Department for Statistics and Sociology of the Republic of Ukraine 2004, p. 310–11).

27 *Nezavisimaia Moldova*, 17 February 2000 (reprinted from *Romania libera*, no. 3007); *Moldavskie vedomost,i*, 4 March 2000.

In 2001 the former special coordinator of the Stability Pact, Bodo Hombach, thus stressed that relations with Romania were of vital interest to Moldova. Indeed, some hold the view that the Stability Pact, where Romania and Bulgaria clearly play a leading role, will become 'a second European Union' once trade has been largely liberalized (Vorob'eva 2004, p. 164).

Moldova's accession to the Stability Pact (SP) offered new possibilities for cooperation between the two countries and for the implementation of projects suspended because of insufficient resources. In May 2001 Moldova joined the SP project on the development of a regional energy market, which offered some $ 1 million for the renovation of the electricity grid and the building of a new high voltage line that would ensure better interaction between the Moldovan and the Romanian energy networks. Within the same framework it was planned to electrify the railway from the Ukrainian border via Chisinau to the Romanian border at Ungheni, to finish the construction of the highway linking Cimislia to Giurgiulesti and to build a motorway bridge between Ungheni and Iasi. The renovation of the bridge across the Prut River at Lipcani-Radauti is currently being implemented. It is financed through PHARE-TACIS funds and will facilitate the movement of goods and passengers in the Upper Prut Euroregion.

Moldova, together with Ukraine, Turkey, Russia, Bulgaria and Romania, was also invited to participate in a project aiming at the creation of pan-European transportation networks and corridors in the Black Sea region. Since Moldova has common borders only with Romania and Ukraine, its participation in this project, which will cost up to € 500 millions, will contribute, above all, to the development of trade and transportation with these countries. The project will improve both rail and highway links, unify security norms, connect traffic controller services and ensure better coordination within the transportation network.[28] The last objective is especially important, since Moldova and Romania use different railway gauges. At present, the frontier between Moldova and Romania has seven crossing points (five of them by road and two by rail).

One of the most serious obstacles to the development of cooperation between Moldova and Romania is their different status with regard to the EU. Romania is a candidate for EU membership, whereas the EU's relations with Moldova are covered by the Partnership and Cooperation Agreement which does not provide a perspective for accession. The two countries thus benefit from different programmes with different levels of support and different rules and procedures, Romania receiving funds through PHARE and Moldova through TACIS. This gap will widen once Romania will be an EU member state and Moldova come under the New Neighbourhood Instrument. Romania currently concentrates its efforts on establishing good relations, including active economic and trade relations, with EU member countries, and on implementing reforms in order to satisfy EU requirements for accession. In this context relations with Moldova, as well as with other neighbours, rank low on the Romanian agenda.

---

28 *Moldavskie vedomost,i,* 27 October 2004.

In early 1997 the Romanian president launched an initiative on cross-border cooperation between Romania, Moldova and Ukraine. The basic treaty between Romania and Ukraine (2 June 1997), *inter alia* included the creation of the Upper Prut and Lower Danube Euroregions. This intention received confirmation at the Izmail summit on 3 and 4 July 1997, when the presidents of the three countries signed the Declaration on Cross-Border Cooperation and their governments a protocol on trilateral cooperation.

The agreements on the creation of the Lower Danube Euroregion along the southern part of the border and of the Upper Pruth along the northern part were signed respectively in Galati (14 August 1998) and in Botosani (22 September 2000) by representatives from Romanian, Moldovan and Ukrainian local administration bodies after a consent had been reached by the presidents of Romania, Moldova and Ukraine. The Council of the Lower Danube Euroregion was composed of representatives from the Cahul (Moldova), Braila, Tulcea and Galati (Romania) and Odessa (Ukraine) regions, and the Upper Prut Euroregion of representatives from the Suceava, Botosani (Romania), Balti and Edineti (Moldova) and Chernovtsy (Ukraine) regions. In October 2002 the Euroregion Siret-Prut-Nistru was created by the Ungheni, Chisinau and Lapusna districts (Moldova) and the Iasi and Vaslui districts (Romania).

These Euroregions were created to provide for new opportunities to solve common problems in the areas of transportation and communication as well as economic development, and to take common action to protect the environment, develop tourism and cooperate in the fields of education, health care and culture. However, up to now, these opportunities have not been used to full extent, above all because political tensions prevalent during previous years did not encourage local administrators to initiate common projects. The legacy of interstate tensions also delayed the signing of agreements between the concerned states which would raise the competence of local governments to participate in the Euroregions' activities. In fact, current regimes regulating the cross-border flow of goods, capitals and labour within these Euroregions are not all that different from those at the international level as cross-border cooperation has been hampered by a triple deficit with regard to economic potential, cooperation and integration.

An important obstacle to boosting cooperation is the predominantly agricultural character of the economy on both sides of the border. A second cause has been the low potential local authorities have for financing common projects. In this respect EU support has been crucial. TACIS has funded centres for cross-border cooperation in the Euroregions, as well as a significant number of common projects between Moldova and Romania, such as the establishment of centres to support small businesses and cross-border trade, business development centres, business information centres and centres for cross-border cooperation which have contributed to a better business climate and allowed the establishing of sustainable contacts between Moldovan and Romanian businessmen.[29] In addition, there has been a significant increase in cultural

---

29 *Moldova Suverana*, 4 March 2004.

**Table 9.2	Implemented and on-going TACIS projects directly contributing to the development of Moldovan-Romanian cooperation**

Project title	Budget (in €)	Status
Radauti–Lipcani border crossing and approach road	2,863,000	on-going
Support to Moldovan export promotion organisation	3,000,000	on-going
Creation of an appropriate decentralised institutional framework to support Moldovan companies' capacities for cross-border and international operations	195,660	on-going
Creation of a Middle Prut National Park based on the National Reservation of Padurea Domneasca	358,909	on-going
Creation of an association of Moldovan border regions and the establishment of a network of public and private sector actors involved in cross-border and regional socioeconomic initiatives	360,640	on-going
SMART – Strategy Development for Moldovan and Romanian Twinnings	297,500	on-going
Leova Youth Information and Documentation Ccentre	172,800	on-going
Traceca infrastructure improvement border crossing	2,000,000	on-going
Border post at Giurgiulesti	1,900,000	on-going
Border management	1,850,000	on-going
Reform and modernisation of the Moldovan customs administration	799,917	on-going
Developing rural tourism (Ecotour)	202,828	completed
Lapusna–Vaslui: 'a bridge for the world'	164, 472	completed
Restoration and maintenance of Manta lakes ecosystem	234,760	completed
Sustainable regional development through creation of agency for CBC (Lapusna/Vaslui)	207,858	completed
Upper Prut – a new tourist market offer for Moldova and Romania	250,298	completed
Business centre of Cahul	262,174	completed
Setting-up of the Lower Danube Euroregion information network	250,100	completed
Centre for SME for support and development (Ungheni)	154,558	completed
Ecological database for the natural reserve Emil Racovita (Upper Prut)	150,086	completed
DECOR – developing cross-border cooperation relationships	202,828	completed
Management plan for biodiversity protection and sustainable development in protected areas of the Lower Danube Euroregion	241,215	completed

Table 9.2 continued

Project title	Budget (in €)	Status
AGROSYS – development of an enterprise start-up and management system for small farmers in Moldova and Romania	228,500	completed
Traceca border crossing terminals	1,511,500	on-going
Common legal basis for transit transportation	1,983,200	on-going
Cross-border cooperation in Ungheni	1,975,000	on-going
Total:	21 305,331	

Source: TACIS Branch Office in Moldova.

contacts and cooperation in the field of education. The universities from the Lower Danube Euroregion, for example, have established an association to facilitate the exchange of professors and students. Finally, TACIS helped implement a number of environmental projects in all three Euroregions, such as the protection of lakes along the Danube on Moldovan and Ukrainian territory in the Lower Danube Euroregion, thanks to a grant of ECU 2,1 million.

The Euroregions' councils and working commissions organize more or less regular meetings during which they discuss projects to be submitted to EU bodies and donor countries for financing. Much remains, however, to be done, such as adjusting the legislation of the three countries, redefining the legal status of the concerned regions, creating regional bodies to audit projects, stabilizing the personal membership of working commissions and establishing free trade zones. The competencies of the Euroregions and the common interests of theirs members are still ill-defined. Indeed, the main prerequisite for the effective functioning of the Euroregions are well-established functional structures and a clear definition of the authorities who are being delegated to the Euroregions by the participating governments.

In the case of Moldova, cross-border cooperation within the Euroregions has thus been hampered by several legal and institutional difficulties. On the one hand, territorial-administrative units in the Republic of Moldova have a much lower economic potential than their counterparts in Romania, a situation which has become worse since an administrative reform in 2003 which has doubled the number of districts included in the Euroregions. On the other hand, while both central and local authorities in Moldova and Romania do not play a very active part in cross-border cooperation, Moldova in addition has no governmental structure to promote the state's regional policy and to defend the Euroregions' interest against the central authorities; Moldovan administrative-territorial units (counties) do not have the capacity to promote individual cooperation policies, no regional integration strategies or specific action plans have been developed either at the central or county levels, and funds from local budgets are insufficient to implement cooperation activities and to gain access to external funding sources.

The Context of Wider Europe and Neighbourhood Programmes

These shortcomings can be partly explained by the fact that the participating institutions have not been able to deal successfully with different systems of financing and project administration. PHARE cross-border cooperation was conceived for cooperation between EU member states as well as between member and candidate states, neither of which applies to Moldova (European Commission 1998). So far, cross-border cooperation on candidate countries' external borders has been financed through national PHARE programmes. During the period 2003–2006 the geographical scope of PHARE-CBC will be extended to cover the external borders of Bulgaria and Romania. In Moldova, CBC programmes continue to be funded through the TACIS-CBC programme (Council of the European Union 1999c). With regard to the financial management of funds, the present system therefore has suffered from mismatched levels of funding, a diverging programming process through separate programming exercises, heterogeneous project selection through separate assessment and external procurement, and different procedures for project monitoring (reporting, monitoring and evaluation) (Commission of the European Communities, 2003b).

For these reasons the EU has, in 2004, launched its New Neighbourhood programmes for cross-border cooperation between EU members and their neighbours. These programmes will

- allow beneficiaries to apply for a single programme, thus harmonizing application procedures and enabling joint decision-making for project selection on both sides of the border;
- enable the allocation of common funds to partners on both sides of the border;
- ensure that no new financial rules or regulations will be required.

In addition, the European Commission has defined four general objectives for the new programme: support for sustainable economic and social development in cross-border regions; cooperation in response to common problems (environment, public health, organized crime, traffic in human beings, migration, etc.); effective and secure borders; promotion of interpersonal (cultural and social) contacts.

After a meeting of experts from Moldova, Romania and the European Commission on 10 February 2004 in Galati (Romania), the Moldovan government has set up a working group to prepare proposals for the Neighborhood Republic of Moldova-Romania for the period 2004–2006. The Neighbourhood programme, in which Moldova and Romania will be equal partners, will focus on the involvement of local and regional governments, the establishment of partnership relations, a strategic approach to the planning process and on the elaboration of an indicative multi-annual plan for financing. It will cover the entire territory of Moldova the and the Romanian *judets* (districts) of Botosani, Iasi, Vaslui and Galati.

Taking into account recommendations made by the European Commission, the Moldovan government, on 15 March 2004, has created two bodies:

- a cooperation council (formed of ten members from the working group) who will coordinate and supervise all activities related to the elaboration of the programme;
- a joint task force (steering committee) who will elaborate a programme document select proposals for the implementation of projects within the framework of the programme.

Since the initiation of the New Neighbourhood Programme by the European Commission, contacts between the governments of Moldova and Romania through their representatives in the cooperation council and the joint task force have become more active and frequent. Despite stagnating political relations between Bucharest and Chisinau throughout 2004, formulation, discussion and finalization of the programme have been speedily achieved, thanks in part to experts from the European Commission. Since then, the European Commission has committed € 27 million (€ 5 million through TACIS-CBC for Moldova and € 22 million through PHARE-CBC for Romania) for the implementation of the programme, in which not only local administrations but also NGOs, educational and research institutions and the private sector will be able to submit joint projects.

A Concept of Cross-Border Cooperation for 2004–2006 within the framework of the programme was approved by the government of Moldova in September 2004.[30] The Concept has made extensive use of the lessons learned from CBC between EU member states and has successfully complied with the Programme on Cross-Border Cooperation for 2004–2006 advanced by the European Commission within the framework of the Wider Europe – New Neighborhood initiative of March 2003. It has also included the main criteria recommended by the European Commission for the selection of projects, that is, that they will have to be implemented by both parties and to take into consideration the tasks outlined by the European Commission, namely support for economic and social development in the border regions, initiation of common actions for environmental protection and health care as well as prevention and repression of organized crime. The Neighbourhood programme on CBC between Moldova and Romania was also included into the EU-Moldova Action Plan. This strategy is viewed by both the Moldovan government and the European Commission as a key condition for the further European integration of the country, as this will ensure the timely implementation of the Action Plan as well

---

30 *Monitorul Oficial* , no. 182-185, 8 October 2004.

as promise a significant improvement in the political, economic, social and cultural relations between Moldova and Romania.

At the same time the political relations between the two countries have visibly improved since the Romanian presidential elections. President Basescu made his first international visit to his Moldovan counterpart Voronin. The official rhetoric of Moldovan state officials on Romania has also changed a lot since Moldova, after the parliamentary elections of March 2005, declared its firm commitment to Europe. During the first half of 2005, Presidents Basescu and Voronin met on several occasions such as the summit of the GUUAM (Georgia, Ukraine, Uzbekistan, Azerbaijan and Moldova) group, a political, economic and strategic alliance designed to strengthen the independence and sovereignty of these former Soviet republics,[31] and the investiture of the president of Ukraine, Yushchenko, and pushed forward the GUUAM agenda which now has a clear political stance. Once the EU became more active in the Transdnistrian conflict and its special representative to Moldova started his mission, Romania, too, began to show more interest in negotiations to solve conflicts and declared once more its availability to provide assistance to Moldova in the European integration process.

Conclusions

Several essays in this book raise prescient questions about the long-term effects of EU policy, and the New Neighbourhood policy in particular, on interstate relations and cross-border cooperation within the context of Wider Europe. There can be no doubt that dynamics of social, economic and political exclusion are in constant tension with the desire of the EU to create a new quality of political community within and without its borders. This conflict is clearly demonstrated by the evolution of interstate relationships between Moldavia, Romania and Ukraine. However, I argue that the EU has, in fact, generated a geopolitical momentum capable not only of transcending historical differences in the region but also of providing improved prospects for cross-border cooperation and economic development. Despite their difficulties, the Euroregions that promote cooperation between Moldova, Romania and Ukraine demonstrate that a learning process of dialogue and accommodation has been enhanced by Romania's candidate status and the opportunities provided by the New Neighbourhood policy. While this assessment might appear optimistic in the face of our politically turbulent times, the last fifteen years give evidence of a positive evolution in interstate relations in very complex and contested political contexts.

---

31 See <www.guuam.org>.

Chapter 10

Euroregions along the Eastern Borders of Hungary: A Question of Scale?

Béla Baranyi

The political changes that have taken place since 1989 have led to a radical transformation of local governance and cross-border cooperation in the border regions of central and eastern Europe. The Carpathian Basin, where Hungary, Poland, Slovakia, Romania and the Ukraine all share borders, is a good case in point to illustrate how increasing cross-border cooperation stimulated by the EU enlargement coexists, and will continue to coexist, with older processes of separation and closure (Ruttkay 1995; Horváth 1998; Nemes Nagy 1998). Since the early 1990s, Hungary has championed policies of more open borders and the development of dynamic cross-border cooperation, particularly between urban centres. Its main reasons have been the existence of sizeable Hungarian minorities in neighbouring countries as well as a more global need to overcome the structural fragmentation of the Carpathian Basin. At the same time, however, Hungary has had to pursue the EU's geopolitical agenda and to ensure an 'adequate' regulation of state borders and border crossings. The eastward shift of the outer limits of the Schengen area but also the tight timetable for lifting restrictions on the labour market and, more generally, mobility mean that Hungary's borders with Romania and the Ukraine will define the outer confines of the EU for some time to come. As a result, the separating role of state borders is likely to cast a shadow on the improvement of Hungarian-Romanian and, to a lesser extent, Hungarian-Ukrainian relations. In all probability the free movement of persons and goods will remain severely restricted, thus exacerbating developmental problems in the border regions which, in socioeconomic and structural terms, are among Europe's most disadvantaged areas. Against this background the question of how to organize cross-border regional cooperation assumes a crucial role.

The main focus of this essay, which is based on research conducted within the framework of EXLINEA (Lines of Exclusion as Arenas of Cooperation: Reconfiguring the External Boundaries of Europe – Policies, Practices, Perceptions), is twofold. On the one hand, it tries to bring into context the significance of cross-border cooperation in terms of geopolitical situations of fragmentation and the peripheral status of border regions. On the other hand, it discusses the possible role of institutionalized forms of cross-border cooperation for the development of these border regions. Here, Euroregions are destined to play a crucial role by offering flexibility in developing territorial frameworks for regional governance transcending both historical and present-day borders. As regional development and support to peripheral areas are major priorities of EU policy, cross-border regional organizations have a vested interest in cooperation. This, at least theoretically, should help promote greater cohesion and lead to the re-establishment of functioning markets in these border regions. One of the principal issues that will be addressed here is that of scale. Cross-border cooperation in the Carpathian Basin has developed at macroregional, regional and local levels, each of these with their own specific characteristics and agendas. The appropriateness of particular scales of action to specific cooperation tasks is thus a major object of scientific investigation into cross-border cooperation dynamics and will be discussed as well.

Geopolitical Contexts, Peripheral Status and Border Regions in Eastern Hungary

From a Hungarian viewpoint the redrawing of state borders in the aftermath of the Versailles peace treaty, and more particularly the Trianon treaty signed on 4 June 1920, resulted in an extreme fragmentation of the Carpathian Basin. Previously interconnected and interdependent communities were separated and promising regional initiatives came to an abrupt end. The new international boundaries between Hungary, Romania and the new state of Czechoslovakia were largely drawn in an arbitrary and artificial manner, neither following ethnic or spatial/structural principles nor taking into account existent economic, social or cultural ties. The consequences of these decisions were far-reaching and continue to affect interstate relations in the Carpathian Basin, even within the present context of a gradual 'Europeanization' of the region. One example of the long-term legacy of the Trianon treaty is the developmental trajectory of the Carpathian Basin. Socioeconomic and spatial structures that had developed organically in the Carpathian Basin rapidly deteriorated and long-standing relations within the region, now criss-crossed by state borders, ceased to exist after 1922. Regional development in the region was also hampered by an atmosphere of interstate hostility and a pervasive lack of trust directly after the first world war (Tóth 1997; Golobics and Tóth 1999; Rechnitzer 1999a, 1999b). Another aspect of this regional fragmentation was the breaking-up of a vast Hungarian cultural space and the subsequent restrictions on interaction between Hungarian nationals and the Hungarian ethnic minorities living beyond Hungary's new borders. In the aftermath of the Trianon treaty and during the two decades following the second world war, formal avenues for cross-border interaction no longer existed. Normal bilateral cross-border traffic only resumed in the 1960s (Sallai 2003). A significant factor in the maintenance of regional fragmentation was the opposition to greater cooperation, voiced both openly and in subtle ways by Hungary's neighbours, as a result of mutual mistrust and fear of Hungary's geopolitical ambitions.

The economic and political marginalization of a significant part of the Great Hungarian Plain has been symptomatic of this geopolitical situation. Hungary's eastern regions, whose economy is largely agricultural with few industrial centres, fell victim to an economic depression that subsequent policies of regional development and state-induced industrialization during socialist times could only partially compensate. The region as well as the larger border areas adjacent to it, which have been the subject of extensive empirical research (Erdősi and Tóth 1998), have been aptly characterized as 'peripheries within the periphery'. At present the gross domestic product of the entire eastern Hungarian border area (which encompasses the 137-km long Hungarian-Ukrainian and a 448-km long stretch of the Hungarian-Romanian border) and of the adjacent Romanian border area lags far behind national averages and the GDP of neighbouring central regions (Ruttkay 1995).[1] In addition, the transboundary region as a whole has a poor transportation infrastructure and is thus insufficiently connected to the larger European economy.

The systemic changes that have taken place in central and eastern Europe since 1989 have created a potential for more successful regional development in the Carpathian Basin, a region greatly burdened by historical, political, socioeconomic and ethnic tensions. It was clearly in the interest of the involved countries to reduce the separating role of state borders and to turn the peripheral location of the border regions into a competitive advantage. However, until now, the regions along Hungary's eastern borders seem to have advanced only slowly in this direction, rarely going beyond the tacit recognition of the necessity of good transboundary relations, and the present state of cooperation leaves much to be desired. Both in quantitative and qualitative terms, the degree of economic, scientific and educational cooperation remains rather low. As in the case of numerous other border regions, cooperation appears to be most successful in the cultural realm (sports activities, city partnerships and student exchange). In many cases the mere existence of ethnic Hungarians in the border areas adjacent to Hungary is one of the main hurdles. Political elites of Hungary's neighbours often continue to perceive closer interaction between these Hungarian minorities and Hungary as a threat to national sovereignty. This, together with the constraints due to the Schengen agreement, is likely to be detrimental to the development of good neighbourly relations as well as to the situation of these Hungarian minorities.

The Emergence of Euroregions in Hungary

Euroregions as an institutionalized form of cross-border cooperation have a long history in Western Europe. Prior to the emergence of these bodies there was no real organizational framework to represent the interests of regions 'divided' by state boundaries. The main objective of Euroregions is to develop mechanisms of border-transcending governance to address economic, environmental, social, cultural and

---

1   Comprehensive empirical studies, which all confirm the peripheral status of the region, were carried out in 1998 and in 2002 in 119 settlements of the Northeastern Great Plain, that is the counties of Hajdú-Bihar and Szabolcs-Szatmar-Bereg (Baranyi,1999, 2001 and 2004; Baranyi et al. 1999).

other regional issues (Scott 1996). Euroregions are expected to help border regions achieve a critical mass of public and private sector activities big enough to strengthen their cohesion and to attract investment by assuming lobby functions (Rechnitzer 1999b; Horváth 2000).

Generally speaking, Euroregions are a product of the postwar rapprochement and integration within Europe. The establishment of the European Common Market in 1957 as well as of other supranational institutions in the years that followed provided an impetus for projects of integration at the subnational and local level. Euroregions gradually appeared in the 1960s but literally boomed in the 1990s with the enlargement process and after EU funds for cross-border projects became available. In central and eastern Europe Euroregions have become increasingly popular, too. There are now well over a hundred regional cooperation agreements, Euroregions and other forms of cross-border cooperation, although less than one-third of them are truly operative.

In a part of the world where tensions, conflicts, non-cooperation and separating borders have for so long been a part of everyday life, the emergence of Euroregions must ultimately be seen as a positive development. And yet it could be argued that cross-border cooperation in the strict sense of the term has only succeeded in the more firmly integrated areas of the EU. As other contributions to this volume indicate, attempts at cross-border cooperation between the recently admitted EU member states and along the EU's external borders, such as between Hungary

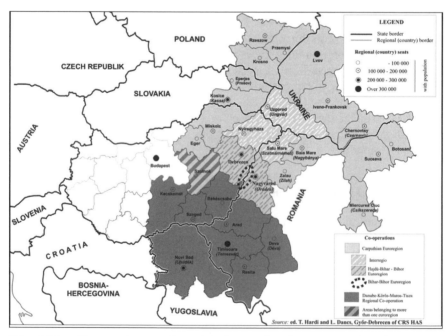

Map 10.1   Euroregions and regional cross-border cooperation in eastern Hungary, 2002

and its neighbours, are beset by a number of problems, among which an excessive number of institutions, a weak organizational basis and often cumbersome territorial constellations in the more recently constituted Euroregions. As a result there is a mismatch between functional border regions and the administrative areas which make up Euroregions. The first of these Euroregions were typically very large, and the Carpathian and Danube-Körös-Maros-Tisza Euroregions have been hampered in their efforts by linguistic, administrative and legal complexity. Partly in recognition of this fact, more recent attempts have tried, or are trying, to establish cooperation at a much smaller scale, such as the Kassa-Miskolc Euroregion, the Interregio Bihar-Bihor and the Hajdú-Bihar-Bihor Euroregion (see Maps 10.1 and 10.2). These will be discussed in more detail below.

One positive factor in the evolution of cross-border cooperation between Hungary and its neighbours has been the gradual but by no means always easy

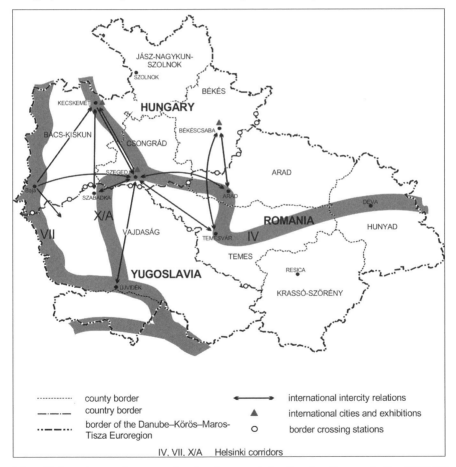

Map 10.2   Axes of economic development in the territory of the DKMT Euroregion

recognition of the mutual interests of microregions and settlements on both sides of the border. This recognition has taken place on the basis of a critical reassessment of the topograpical, structural and ethnic interdependence of territories divided by the Trianon treaty. At the same time, these new Euroregions attempt to avoid political controversy by circumventing obstacles at the national level. As appropriately stated by one researcher, the importance of these incipient interregional structures lies in the fact that they provide, through cross-border relations, a 'political content and context that contributes to the establishment of new European spatial dimensions (Scott 1997).

The Case of Large-Scale Euroregions

The Carpathian Euroregion Interregional Alliance, or Carpathian Euroregion, was established on 14 February 1993 as the first of its kind in central or eastern Europe. It is unique in a number of ways. It covers not only adjacent counties but other medium-level administrative areas in five countries, namely Hungary, Poland, Romania, Slovakia and Ukraine. In this respect it resembles the Alpine-Adriatic Working Community created in 1978, which has indeed served as a model.[2] Secondly, it was established along what were then the EU's external borders with former socialist states. Thirdly, it corresponds to a macroregion which has a peripheral status not only within Europe but also with regard to all the states involved. In other words, the Carpathian Euroregion is an assembly of socioeconomic peripheries (Illés 1997; Gorzelak 1998; Rechnitzer 1999b).

The second-largest Euroregion is the Danube-Körös-Maros-Tisza Euroregion established in Szeged on 1 November 1997. It attempts to provide an institutional organizational framework for cooperation initiatives in the Hungarian-Romanian and Hungarian-Serbian border regions and covers an area equal to about two third of Hungary involving several spatial units. It can be considered as a multiregional organization similar to the Alpine-Adriatic Working Community and the Carpathian Euroregion. Its importance derives from the fact that it includes areas of Yugoslavia (Rechnitzer 1999b).

Assessments of the efficiency of these two large Euroregions are as yet rather ambiguous. The most problematic case is that of the Carpathian Euroregion, due to its sheer size and its multiregional transnational character. It encompasses an area of 161,000 km^2 and a population of 16 million people, roughly the equivalent of a medium-sized country. Communication between the participating actors is hampered by very different and asymmetric forms of regional and local administrations. Considerable research has been conducted on its operational difficulties (Illés 1993; Baranyi 2002a, 2002b and 2003), and most assessments confirm that disagreement at the level of national politics are frustrating local ambitions for more positive cooperation initiatives (Éger 2000). Although the Danube-Körös-Maros-Tisza

---

2   The Alpine-Adriatic Working Community is made up of regions from seven countries: Austria, Croatia, Germany, Hungary, Italy, Slovenia and Switzerland (www.alpeadria.org).

Euroregion covers a large territory (77,000 km^2) as well and is home to a substantial population (6 million people), its size is considered more 'manageable'. More importantly, it is comprised of four Hungarian counties (Jász-Nagykun-Szolnok, Csongrád, Bács-Kiskun and Békés), four Romanian regions (Krassó-Szörény, Temes, Arad and Hunyad) and the Voidvodina province of Yugoslavia which were all part of Hungary until the Trianon treaty of 1920. Besides a similar physical geography the region is also characterized by a more homogeneous socioeconomic situation and ethnic composition.

A common feature of these two macroregional organizations is their inability to promote cooperation activities in the same measure as Euroregions in western Europe. Top-heavy in state representation, they are also less visible and less socially embedded than their western counterparts. Surveys carried out in the border area of the Northeastern Great Plains in 1998 and in nine settlements on both sides of the Hungarian-Romanian and Hungarian-Ukrainian borders in 2001 and 2002, have shown that the role, objectives and mission of these Euroregions are scarcely known to their inhabitants.

Moreover, in recent years, the peripheral social and economic status of the territories which make up the Carpathian Euroregion has been further reinforced. The lack of resources for rapid development and the resulting economic crisis have led to rising ethnic tensions. Under these circumstances, it is hardly surprising that little tangible results have been achieved. The Carpathian Euroregion has been unable to navigate successfully between the complex political contexts that block its effective operation. Member regions have been reluctant to participate in common activities, and several of them have even considered to leave the organization. The Hungarian county of Hajdú-Bihar thus temporarily suspended its membership in 2004. To sum up, the Carpathian Euroregion appears to have been a step or two ahead of its time. It was established during a period when the consolidation of the new democracies was the principal task at hand and the development of meaningful cross-border relations was not on the top of the political agenda. This is perhaps the reason why the model of a Euroregion based on earlier experience in western Europe proved incompatible. The necessary economic, political and social conditions for institutionalized cross-border cooperation simply did not exist.

The challenge facing the Danube-Körös-Maros-Tisza Euroregion has been to promote cooperation along the Hungarian-Romanian-Yugoslav border, an area of great political tensions, but centred on the historical Bácska and Bánát regions, which are rather homogeneous from a geographical point of view. At the end of the nineteenth century, a period of accelerated capitalist development, these two regions had embarked on a special developmental path. Economic growth was based on a division of labour between large urban centres and on competing regional cities, namely Temesvár, Szeged and Arad. The ensuing social transformation boosted overall regional development and also proved greatly beneficial to smaller centres within the region. Besides possessing a strong and innovative agricultural sector, the region was at the cutting edge in various manufacturing and services sectors and benefited from an excellent municipal and regional infrastructure. The Délvidék

(Southern Country) thus began to emerge as a functional space and economic area. This development came to an abrupt end with the partition of Hungary, after the first world war.

Because of its large size, the Danube-Körös-Maros-Tisza Euroregion can be defined as a multiregional organization similar to the Alpine-Adriatic Working Community and the Carpathian Euroregion, but its stated objectives require a more active and concrete participation from the contracting parties (Rechnitzer 1999b). The managers of the Danube-Körös-Maros-Tisza Euroregion seem to have learned from the problems obstructing cooperation in the Carpathian Euroregion and are very keen on avoiding the pitfalls of 'macropolitics'; recent anti-Hungarian incidents in Temesvár and the Voidvodana have served as warning signs. Indeed, the strategy employed has been one of focusing only on the development of local and regional relations, otherwise maintaining a low profile. As a result, economic and social relations at the local level have improved, especially along the Hungarian-Romanian border (Éger 2000). Since the extension of the PHARE cross-border cooperation programme to this area in 1996, a regional coordination office has been set up. Nevertheless, achievements so far have not yet met expectations (Terra Studio 2000). Real progress will only come with a certain degree of institutionalization of cross-border decision-making processes. It is already evident that member regions do not participate on an equal basis. Most active are the regional networks of Békés (Hungary) and Arad (Romania) and of Csongrád (Hungary), Temes (Romania) and Voidvodina (Serbia). These networks have also been more successful in promoting

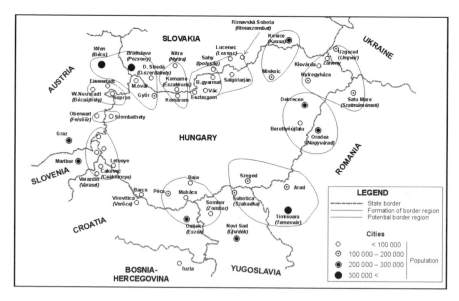

Map 10.3   Potential and developing border regions and interregional cooperation initiatives

their interests, especially through lobbying for support of large-scale infrastructure projects, while other member regions have favoured small projects. This has, however, led to a kind of stalemate, as no single project approved by the Danube-Körös-Maros-Tisza Euroregion during the years 1998–2000 has enjoyed unanimous support. In addition to these operational problems, the Hungarian county of Jász-Nagykun-Szolnok suspended its membership in the Euroregion in 2003.

Toward Smaller Euroregions?

Alternatives to these large and ineffective cooperation structures are developing from interurban relationships driven by the economic and political logic of EU enlargement. Several urban networks and subregional cooperation initiatives appear indeed to promise better prospects for the implementation of projects and for regional development. More flexible and closer to the citizens they might also be able to promote a better sense of common cross-border identity while at the same time supporting efforts to convey more meaning to macroregional cooperation. In eastern Hungary for example, such cooperation is taking shape between urban centres in three border regions: two of these networks have already been established between the Hungarian city of Nyíregyháza and the Ukrainian city of Ungvár (Uzhorod) and between the Debrecen and Berettyóújfalu in Hungary and Nagyvárad (Oradea) in Romania; a third Hungarian-Romanian-Serb network involving the cities of Szeged, Arad, Temesvár (Timişoara) and Szabadka (Subotica) is under consideration.

In addition to these urban networks there exist several interregional associations and Euroregions of a smaller scale, and others are in the process of being formalized. Along the Hungarian-Slovakian border for example, the Kassa (Košice)–Miskolc Euroregion has been created on the basis of improved political and economic relations between the cities of Miskolc (Hungary) and Košice (Slovakia). An Interregio, in the form of a trilateral cooperation project between the Hungarian county of Szabolcs-Szatmár-Bereg, the Romanian county of Szatmár and the Ukrainian county of Transcarpathia has been established on 6 October 2000(see Map 10.1). Its aim is to develop cooperation between local governments within the larger framework of the Carpathian Euroregion, among others through implementing objectives defined by Hungarian-Ukrainian, Hungarian-Romanian and Romanian-Ukrainian bilateral treaties in the fields of environmental protection, management of water resources, economic development, the comprehensive development of infrastructure and tourism, education and training as well as the management of interethnic relations and a common cultural heritage. The formalization of this Interregio, whose institutional basis will be the urban network of Nyíregyháza-Ungvár-Szatmárnémeti, into a full-fledged Euroregion is in progress.

Among the most recent cross-border organizations along the Hungarian-Romanian border are 'microregional' Euroregions and Euroregions at county-level established in 2001 and 2002 (see Map 10.4). An example of the first is the Bihar-Bihor Euroregion centred on the small cities of Biharkeresztes (Hungary) and Bors

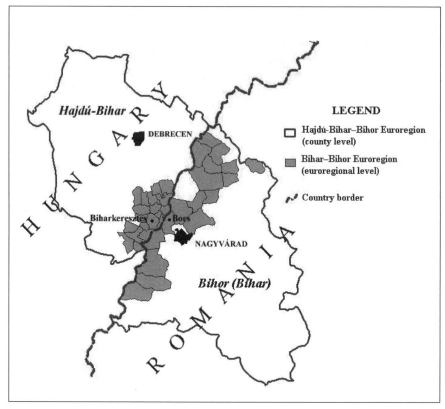

Map 10.4   The   Hajdu-Bihar-Bihor   (county-level)   and   Bihar-Bihor (microregional) Euroregions

(Romania), an example of the second the Hajdú-Bihar-Bihor Euroregion. These Euroregions, which cover an area of almost 14,000 km^2 and have a population of about 1.2 million people, 're-integrate' an historic functional regional and settlement area and are assisted by the Regional Development Associations of Bihar Border Settlements created in 1998 on both sides of the border. Both Euroregions are open to any municipality, local government association, organization emanating from civil society or other legal entity with a community development agenda, provided it is registered in the Romanian county of Bihor, the Hungarian county of Hajdú-Bihar or a territory covered by one of the cross-border microregional associations. Besides accepting the statutes of these Euroregions, members must actively participate in the implementation of its objectives. Both Euroregions adhere to standards of cross-border cooperation and international relations recommended by the European Union and established by Hungarian-Romanian interstate agreements. Their goals are very similar. Among their prime objectives are the promotion of socioeconomic development, the implementation of strategies of sustainable development and local participation in and management of European programmes. Their statutes also emphasize the importance of direct bilateral relations for cross-border relations while preserving strategic elements of the 'macroregional model'. A so-called working group of the Bihar-Bihor Euroregion is responsible for promoting cooperation between municipalities of the border settlement associations in the fields of economic development, tourism and culture. The Hajdú-Bihar-Bihor Euroregion, centred on the Hungarian city of Debrecen, is pursuing similar objectives through cooperation between larger, medium-level spatial units. Cooperation within the former historic county of Bihar appears to have a better chance of succeeding where other cross-border organizations have failed: it can look back on a tradition of economic integration, urban infrastructure and other often organically interdependent functions that was interrupted by artificially created borders in 1920. The Romanian (and formerly Hungarian) city of Oradea (Nagyvárad), once the economic centre of the region for centuries, has thus become once more a logical centre of attraction.

Another feature shared by small Euroregions in eastern Hungary is that their formal establishment was preceded by the creation of cross-border business zones, such as the Zahony Regional Business Zone established by governmental decree in 1996 and the first of its kind in Hungary. The objective of this agency has been to promote the development of the Hungarian-Ukrainian border region through attracting investors. This was followed in 1999 by the creation of the Bihar Business Zone, organized along an axis linking Berettyóújfalu, Biharkeresztes and Nagyvárad. The Békés County and Makó Regional Business Zones have similar aims. Together with the already mentioned Interregio and Euroregions of Hajdú-Bihar-Bihor and Bihar-Bihor these business zones have been able to expand cultural and educational cross-border relations and to obtain significant achievements in economic cooperation, despite a chronic shortage of funds. However, one should not expect rapid growth or even an economic 'turnaround' in the short- to mid-term.

Other Potential Levels of Interregional Cooperation

As early as the mid-1990s Hungarian scholars of regional studies have placed special emphasis on the role of large cities in border regions for national and international development. Analyses of Austrian-Hungarian cross-border cooperation and development processes in West-Transdanubia have led to the idea that Euroregions possess a considerable potential for (re)integrating fragmented urban networks and their hinterland. This assumption also seems to apply to the more recent Euroregions emerging along the eastern borders of Hungary. It is, however, evident that this potential can only be realized if participating municipalities and regions are able to identify common interests in developing cooperation. This cannot be achieved by the mere institutionalization of cross-border cooperation; local governments, medium-level institutions and organizations emanating from the civil society must also take an active and supportive part. Findings of surveys to date indeed suggest that microregional alliances, municipal associations, urban networks and, above all, large regional centres have a dominant role to play in cross-border regional cooperation, and perhaps even in the support of larger Euroregions. All along Hungary's borders we can identify truncated urban regions of similar size and with complementary functions on both sides of the border (see map 10.3). The organization of these urban regions into cross-border networks by giving priority to the renewal of their infrastructure is possible and would considerably boost cross-border cooperation (Rechnitzer 1999b). In eastern Hungary this would amount to the development of flexible institutional relations between the regional centres of Nyíegyháza, Debrecen and Szeged in Hungary, Ungvár in Ukraine and Szatmárnémeti, Nagyvárad, Arad and Temesvár in Romania, and, further south, between the Hungarian cities of Szeged and Békéscsaba and the major regional centres of Arad and Temesvár in Romania. The key role of large cities and other urban centres for cross-border relations is also demonstrated by trends in the international relations of towns located in Hungary's Southern Great Plain Region. Here settlements in the direct vicinity of the border have established dynamic relations with their neighbours on the basis of mutuality and reciprocity. These towns play a catalytic role in cross-border cooperation and therefore occupy a central position within the Danube-Körös-Maros-Tisza Euroregion.

Conclusions

Since Hungary's accession to the European Union in May 2004 its eastern regions are in a position to break out of their former isolation and become both major centres for the East-West trade and important links for the spreading of socioeconomic and technological innovation. This is, of course, predicated on the condition that Hungary's eastern border regions assert a genuine *connective power* with their Romanian counterparts, leading to almost permeable borders. Despite the problems they are facing, Euroregions still promise to be the most effective forms of cross-border cooperation aiming at regional development. In recent years there has, however, been a movement away from the large Euroregions created in the early 1990s towards cooperation agreements and networks of a smaller scale. Euroregions such as the Carpathian and the Danube-Körös-Maros-Tisza Euroregions have indeed been beset by numerous problems: sheer size, linguistic complexity, a cumbersome institutional framework, a mismatch of partners who often remained under the tutelage of their state government, a top-down approach, insufficiently defined common interests, all contributed to hamper efficient cross-border cooperation and in some cases even led participating bodies to suspend their membership. Urban cross-border networks and subregional cooperation initiatives, on the contrary, do not seem to suffer from these handicaps. By keeping a low profile they have been less subject to pressure arising from interstate tensions. They are more flexible and closer to their citizens, have more clearly defined objectives and are more committed to their implementation. Many of them have been able to build on earlier cooperation agreements such as cross-border business zones established in the early 1990s. More importantly perhaps, some of them, such as the Euroregions centred on Bihar county, try to (re)integrate microregions along historical lines, looking back towards the period prior to the first world war, when the region experienced an economic boom and constituted a functional economic, cultural and social space. It is still too early to tell whether these more recent attempts will succeed in overcoming the peripheral status of the border regions along Hungary's eastern borders in the long run. Achievements to date have been rather modest, partly because of insufficient funds. But for the time being cross-border cooperation based on urban networks and microregions appear to offer the best prospects for regional development initiatives, not least because they might be able to offer a firmer ground for large-scale cooperation, too (Golobics 1996; Rechnitzer 1991a).

Chapter 11

Transboundary Interaction in the Hungarian-Romanian Border Region: A Local View

Gyula Szabó and Gábor Koncz

Much research on Hungarian border regions has focused on their peripheral status, which is due to structural problems, poor infrastructure, lack of investment, etc. (Horváth 1998; Rechnitzer 1999b). Béla Baranyi (see this book) thus describes the development of cross-border cooperation institutions in eastern Hungary as a strategy for overall regional development. As Baranyi illustrates, cooperation models developed in western Europe, such as Euroregions, can only be partially applied to the new EU member states and their neighbours. This essay deals with an issue that has received rather less attention, namely that of local perceptions of the border and cross-border cooperation on Hungary's eastern borders.[1] More specifically, we will discuss how the border affects personal relations between citizens on both sides of the border and to what extent Euroregions and other forms of cross-border cooperation are believed to shape everyday life. The basis for this analysis is a series of largely quantitative surveys carried out in five pairs of settlements located on both sides of the Hungarian-Romanian border (Vállaj and Csanálos, Létavértes and Székélyhíd, Biharkeresztes-Ártánd and Bors, Elek and Ottlaka, and Kiszombor and Naygcsanád) in July and August 2002, before Hungary's accession to the European Union (see Map 11.1).[2] As such, it captures local perceptions during a period when the effects of the impending EU enlargement were still subject to much speculation; inhabitants of the concerned border regions frequently voiced fears of illegal immigration and were apprehensive of the separating affects of the future Schengen borders.

Research was carried out in two stages. During the first stage, mayors and other representatives of local governments were asked about their opinion on local, regional and national cooperation projects, their experience of Hungarian-

---

1   Here we would like to acknowledge the work done by Ulrike Meinhof (2002) and other researchers who, within the framework of the EU research project 'EU Border Identities', have compiled and compared local border narratives.

2   Research on the project 'Borderland Situations and Peripherality on the Hungarian-Romanian State Border' was conducted by the Debrecen Department of the Alföld Research Institute and received support from the Hungarian National Research Fund (OTKA).

Map 11.1   Settlements studied during the survey

Romanian cross-border cooperation and their attitudes towards Romanian partners, their experience and opinion of EU-funded projects, their opinion of self-organized settlement associations and special economic zones and, finally, their view of the consequences of Hungary's EU accession and the implementation of the Schengen agreement on cross-border cooperation. A second survey involving 600 residents of the border region was based on a questionnaire whose main aim was to elucidate perceptual dimensions of 'neighbourhood', existing practical cross-border working relations and links, views on the borderland situation and attitudes associated with it, local perceptions of the potential for regional cooperation and knowledge of regional organizations as well as local demographic contexts. The results of the survey were then weighted for response rates and gender distribution in relation to the total population. However, as the sample cannot be considered representative, the results of the survey can only indicate general trends.

K-means cluster analysis on the basis of demographic indicators (age, profession, educational background, family size, length of residence) led to the identification of four main groups: 'young employees-recent settlers' constituted the youngest group of the sample and accounted for 34.6 per cent; 'little-educated pensioners' made up 25.4 per cent; 'middle-aged entrepreneurs' – mostly male (70 per cent), heads of large families, long-time residents and, generally, with little formal education – accounted for 13.5 per cent, and 'well-educated middle-aged interviewees' – mostly long-time residents, nearly half of them (former) civil servants and one third of them with a college or university degree – for 22.5 per cent of the sample.

Attitudes toward Local Cross-Border Cooperation Initiatives

At the time of the survey, few settlements were involved in cross-border cooperation projects, even though some progress had been made in the field of cultural exchange programmes and more traditional forms of cross-border activities, as other researchers have pointed out (Baranyi 2005). Most prominent among these were meetings and events organized by ethnic groups living in the border region, including activities such as folk dances and mutual visits during national holidays. Other examples were sports events, namely football matches, and student camps as well as official receptions in the form of large banquets. While respondents to the questionnaire were very proud of these, in their opinion all too rare cultural and sport activities, they were highly critical of 'white–tablecloth dinner parties' considered to be a 'waste of resources'. Even local government officials such as the mayors of Létavértes and Biharkeresztes saw little sense in organizing such events.

Among the examples of economic cooperation mentioned by the respondents were a cross-border incubator for local enterprises, established in the Hungarian city of Makó and its Romanian counterpart, economic resource surveys carried out in border settlements by several foundations whose aim it is to develop local enterprises, regular meetings organized by the local chambers of commerce in Biharkeresztes (Hungary), as well as plans for the joint treatment of waste water in Vállaj (see Map 11.1).

Overall, it was felt that the poor quality of relations between the Hungarian and Romanian governments exercised a negative influence. To this can be added differences in the two administrative systems. In Székelyhíd, for example, there appears to be an ill-defined division of labour between the mayor and the local council, and the election of seven mayors during three electoral cycles was thought to make planning of cooperation activities virtually impossible. One of the mayors interviewed also mentioned that Romanian local politicians were often perceived as 'unreliable as Romanian national politics', uninterested in preparing long-term concepts and mainly concerned by personal material gains.[3]

---

3 It must be pointed out that Hungary and Romania have a long history of hostile relations and mutual mistrust. While prejudices and resentment still affect perceptions of the neighbouring country, there are indications that this is changing. Local entrepreneurs, for example, declare that mistrust has decreased considerably since the 1990s and that political conflicts between the two nations are now becoming rare.

Individual Cross-Border Relations

One of the aims of the survey was to assess the character, quality and intensity of cross-border relations for residents of the area. Here, almost 40 per cent of the respondents indicated that they had contacts with Romania (see Table 11.1). The fact that more male (over 44 per cent) than female residents (just over 34 per cent) were crossing the border can probably be explained by the importance of 'petrol tourism'. Members of all four groups were more likely to visit Romania if they had ethnic Hungarians as friends or relatives; in this context the proximity of a border crossing point does not appear to make much difference. Members of the four groups markedly differed as to the reason of their cross-border visits. Members of the first group ('young employees–new settlers') mostly cross into Romania for recreation and work, while those of the second group ('little-educated pensioners') are generally visiting relatives living on the other side of the border. Unsurprisingly, middle-aged entrepreneurs typically indicated contacts with Romanian officials and business partners as their main motive.

The frequency of cross-border visits is closely related to the proximity of a crossing station. Nearly 10 per cent of residents living near such a station are visiting Romania at least once a month, mostly for shopping. Most of them are members of the first and third group. In Létavértes and Kiszombor, where respondents were also less knowledgeable about the names of neighbouring settlements (see below), about two-thirds of the persons questioned had no ties with residents of neighbouring communities in Romania; the others declared making visits to relatives, but overall the intensity of cross-border contacts is low.

Awareness of the name of neighbouring settlements in Romania can be seen as a good though trivial indicator of cross-border relations. There are, indeed, significant differences between the various municipalities as well as between members of the four groups. 'Well-educated middle-aged' respondents mostly gave correct answers, whereas the highest number of incorrect answers came from the group of 'little-educated pensioners'; knowledge of these names was weakest among members of the first group (see Table 11.2).

Table 11.1   Frequency of cross-border contacts (percentage of answers)

	daily	weekly	monthly	occasionally	never
Settlement with border crossing station	1.6	3.2	6.5	59.5	29.1
Settlement without border crossing station	0.6	0.6	2.3	62.8	33.8

Source: survey data.

Table 11.2 **What is the name of the neighbouring settlement in Romania? (percentage of answers)**

	don't know	correct answer	incorrect answer
Biharkeresztes	9.6	79.1	11.3
Ártánd	6.7	93.3	0.0
Elek	13,0	67.8	19.2
Létavértes	12.3	64.8	22.9
Kiszombor	17.9	54.7	27.4
Vállaj	0.0	89.7	10.3

Source: survey data.

Local Perceptions of Euroregions

A major aim of the two surveys was to study local perceptions of the Carpathian and the Danube-Körös-Maros-Tisza Euroregions (for more details on these Euroregions, see Baranyi's contribution). Here, perhaps the most interesting result is the evaluation of the activities of these Euroregions by local government officials interviewed during the first survey. Most of these officials agree that the two Euroregions lack efficiency in their operations and that their activities are too limited, being mainly confined to the occasional distribution of printed materials and small funds. One delegate of the Kiszombor microregion, for example, called the Danube-Körös-Maros-Tisza Euroregion a 'dead organization' without clear functions and without any impact on the everyday life of citizens. The mayor of Biharkeresztes, one of the founders of a rival organization of the Carpathian Euroregion, thought that the latter as well as the Hajdú-Bihar county government were reluctant to support local initiatives, mainly because the county authorities consider the Euroregion unable to take into account local requirements. The mayors of Vállaj and Létavértes expressed no clear-cut opinion for lack of information. Although the former maintained that cooperation in the form of Euroregions was useful and even necessary for EU candidate countries, he was not able to indicate any practical achievements. Similarly, the mayor of Elek, despite his generally positive opinion of the Danube-Körös-Maros-Tisza Euroregion, was at pains to list any significant activities besides the improvement of some cultural relations, the organization of conferences and the publication of a number of papers. However, all these respondents blamed the Romanian partners for the difficulties and problems rather than the Hungarian government or the Euroregions themselves.

During the second survey, local residents were asked if they knew to which Euroregion their settlement belonged and in which way it could contribute to cross-border cooperation. Slightly more than half (50.3 per cent) of the respondents could not answer the first question and 0.6 per cent gave no response. In this respect there was no difference between the residents of the Danube-Körös-Maros-Tisza and the

Carpathian Euroregions. Positive answers were highest (61.6 per cent) among the fourth group ('well-educated middle-aged' respondents), accounted for more than half the responses within the groups of 'young employees–new settlers' (52 per cent) and 'middle-aged entrepreneurs (51.4 per cent), and were lowest for the group of 'little-educated pensioners' (33.1 per cent). Nearly two thirds of the respondents (62.8 per cent) were unable to say whether their county government belonged to a Euroregion and 5.1 per cent gave an incorrect answer. Among those who were aware of the existence of a Euroregion, 55.6 per cent answered correctly ('yes'), 8.3 per cent incorrectly ('no') and 36.1 per cent declared not to know. The majority of correct answers were given by respondents from the fourth and first group (respectively 38 and 35.7 per cent). There was no significant difference between respondents from different settlements. Awareness of the targets and missions defined by the Euroregion the respondent's settlement belong to was low, with 66.1 per cent professing themselves ignorant and 27.1 per cent having only little information; only 1.8 per cent proved adequately or fully aware of these objectives, foremost among them the groups of 'well-educated middle-aged' respondents and of 'young employees–new settlers'.

Cooperation with a neighbouring settlement in Romania was thought possible by 89.6 per cent of the respondents. Those who had answered in the affirmative were then asked how their settlement could contribute to cross-border cooperation with a neighbouring settlement (see Table 11.3).

Table 11.3   **How can your settlement contribute to the establishment of cross-border cooperation with the neighbouring settlement? (percentage of first answers received)**

	NA	Change of experiences	Economic connections	Cultural pro-grammes	Financial aid, donation	Develop-ment of railway and road connections	Opening of borders	Other	Did their best/It works already
young employees-new settlers	9,5	0,8	6	*26,7*	3,5	4,4	14,2	5,1	29,8
old-age pensioners with low education	*16,4*	1,1	5,8	17,3	1,4	*5,4*	13,7	6,4	32,5
middle-aged entrepreneurs	15	0	4,9	11,1	3,5	1,6	*19,8*	*12,8*	31,3
well-educated middle-aged	9,6	1,2	*16,7*	14,2	0	1,2	11,7	3	*42,4*

The cells with the values of adj. Standardized residuals 2 or higher were considered as very typical.

While members of the first group suggested cultural programmes, 'middle-aged' entrepreneurs' advocated more open borders, and the 'well-educated middle-aged' respondents the development of economic relations. Overall, respondents, and particularly 'well-educated middle-aged' respondents, thought that the Hungarian authorities had done their best to improve relations and that these were working fine. Answers varied, however, between respondents from different settlements, who indicated different priorities (see Table 11.4).

Table 11.4 **How can your settlement contribute to the establishment of cross-border cooperation with the neighbouring settlement? (most frequent answers received)**

	Proposed activities
Biharkeresztes	'Did their best'/'it works already'; others
Ártánd	'Did their best/'it works already'
	Change of experiences; economic and
	commercial connections; development of
Elek	*railway and road connections; opening of borders*
Létavértes	*Opening of borders; 'did their best'/'it works already'*
Kiszombor	*Economic and commercial connections; organization of*
	joint programmes ('village days', culture, sports)
Vállaj	'Did their best'/'it works already'

Source: survey data.

Typically, residents from settlements without a border crossing station were strongly in favour of measures to further open the border. This was for example the case in Létavértes where local citizens have been very vocal about the opening of a new crossing station during the years preceding the survey. Similarly the general dissatisfaction of residents from Elek can probably be explained by the fact that there already exists a major crossing station in the neighbouring city of Gyula and that there is little hope for another one at Elek. Conversely, residents from Kiszombor, which already has a crossing station, declared themselves in favour of further developing exiting relations and of improving cultural relations.

Conclusions

The two surveys carried out in mid–2002 provide a snapshot of local perceptions of cross-border relations along the Hungarian-Romanian border. The opening of this border in 1989 clearly led to an intensification of individual cross-border contacts in the form of visits, cultural and sports events as well as business contacts. The same cannot be said for institutional forms of cooperation. The two vast Euroregions established along Hungary's eastern border so far have achieved little, and local government officials have shown themselves highly critical of the workings of these organizations, insisting on their inability to implement projects of regional development that would have an impact on the everyday life of citizens. The main reasons given are differences between the two administrative systems and the inertia of Romanian local politicians. Although a large majority of the residents of the border region are in favour of more cross-border cooperation, many appear to be content with the present state of things. Relatively few appear to be aware of the two existing Euroregions and still less of their workings and their objectives. As Baranyi points out in his contribution to this volume, this situation may have changed during the last years, and it is to be hoped that the recent creation of smaller Euroregions will offer better prospects for cross-border cooperation initiatives that are efficient and play a greater role in regional development.

Chapter 12

Patterns of Legal and Illegal Employment of Foreigners along the Hungarian-Ukrainian Border

István Balcsók and László Dancs

Processes of increasing economic integration and political cooperation initiated during the postsocialist transition period have greatly affected the significance of state borders in central and eastern Europe. More recently, the historic EU enlargement of 2004 has eliminated most of what remained of formerly restrictive border regimes within the 'new' Europe and is now transforming those along the new external borders (see, for example, Mrinska, Popescu and Baranyi in this book). As several contributors to this volume stress, EU enlargement and its (re)bordering effects are characterized by simultaneous processes of inclusion and exclusion. Nowhere is this contradiction more pronounced than in the issue of migration and labour mobility. In the 'old' member states in particular, widespread popular fear of job losses and public insecurity caused by open borders in the east have found their way into political discourse in the years leading up to accession and still remain vivid. Many of these countries have introduced quotas to limit labour migration, and strict visa regulations apply for citizens of non-EU countries such as Ukraine and Russia, even though recent studies have shown large-scale labour migration to be highly unlikely.

In a similar way a large number of representatives from local governments in northeastern Hungary, and the Szabolcs-Szatmár-Bereg county in particular, indicate immigration and foreign labour (primarily from Romania and Ukraine) as one of the major problems of their region, most likely reflecting popular sentiment. Against the backdrop of heavily politicized debates on Hungary's future role within the EU and potential benefits and disadvantages of EU membership, the spectre of illegal immigration and employment was raised by national and regional media. Much of this debate reflected the polarized political landscape that emerged after the mid– 1990s and was rather emotional than rational.

This paper will briefly outline research carried out between 1998 and 2002, in the years preceding Hungarian adoption of EU norms regarding visas and immigration. The following is an attempt, based on official data and surveys conducted in the border area, to study labour migration along the Hungarian-Ukrainian border and to show that its role in the local economy is far from that ascribed to it by popular prejudice. What surfaces is not only a contradiction between reality and perceptions of a threatening 'other' but also incongruencies between EU and national policies and local economic contexts.

Trends in Cross-Border Traffic

The intensity of pedestrian and vehicular traffic at Hungarian border stations has varied considerably since the opening of the border in 1989, subject to a number of political shocks and changing opportunities for trade.[1] At Záhony, then the only crossing point on the Ukrainian-Hungarian border, over two million crossings were registered in 1988, and over 1 million in 1989, most of these by transit passengers between the Soviet Union and Yugoslavia who tried to profit from bartering and petty trade as well as different exchange rates. After political relations thawed and four new crossing points were opened, traffic increased dramatically until the collapse of the Soviet Union in 1991 when the rouble lost much of its former value and Ukraine became independent: cross-border traffic ceased and transit travel was reduced to a trickle. Subsequently traffic picked up once more, peaking at more than 15 million crossings in 1995, but fell after Ukraine restricted imports through high customs duties.[2] Traffic increased again in 2000 until the introduction in August 2003 of stricter visa regulations for Ukrainians in preparation of Hungary's EU accession led to a drop in the number of crossings and resulted in the temporary closure of two smaller crossing points at Barabás and Lonya.

Foreign Workers in the Hungarian Labour Market

To protect the domestic labour market the Hungarian government has passed legislation which does not allow the employment of a foreigner if a Hungarian citizen can fill the job. This legislation also limits the number of work permits to 81,000 per year, despite a rising number of applications since the 1990s.[3] During the first six months of 2002 the majority of work permits were issued to nationals of Romania (60 per cent), Ukraine (13 per cent) and Slovakia (6 per cent) as well

---

1   Rail traffic decreased substantially after 1988 when international trade became much less important.

2   The period from 1992 to 1995 is popularly called 'Z-tourism' because it was possible to export duty-free second-hand cars of Russian or Ukrainian make by using temporary Hungarian licence plates marked with the letter Z and to reclaim value-added tax for these vehicles. Since 1999 petty trade has almost disappeared, except for 'petrol tourism' taking advantage of lower prices in Ukraine.

3   It is of course impossible to determine the precise number of foreign workers in Hungary as firms are sometimes operating illegally and many workers are employed without the required permits. In addition, there are certain loopholes such as the creation of a private company whose managing director need not apply for a work permit – a procedure exploited notably by Chinese nationals.

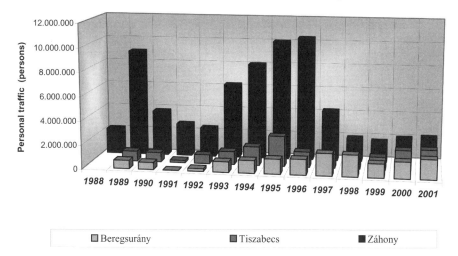

Figure 12.1   The development of rail and road traffic at crossings on the Hungarian-Ukrainian border, 1988–2002

as the People's Republic of China; most of these arrivals were ethnic Hungarians from neighbouring countries whose integration into local labour markets has been relatively easy.[4] And whereas an increase of 13.4 per cent could be observed for Romanian nationals in 2002, these represented only 1,645 workers. The employment rate of foreigners in Budapest and the neighbouring Pest county, where almost three quarters of the foreign workers are employed (see map below), seems indeed to have reached its peak in 2002. When compared to countries in western Europe, the share of officially registered foreign workers is thus quite low, thanks in part to rigid labour legislation.

The Labour Market in the Northeastern Region

According to a survey undertaken in 1998 by the Hungarian Social Research Institute (TÁRKI), foreigners were residing in over half the municipalities (57 per cent) located in the Northeastern Region, but accounted for only 0.4 per cent of the total population (Sík 2002).[5] Data from local governments show that the majority of them were legal immigrants (38 per cent) or ethnic Hungarians from abroad (23 per cent), followed by foreign workers with temporary residence status (11 per cent), illegal workers (6 per cent) and refugees (2 per cent), the latter three all highly concentrated at certain locations within the region.

---

4   Nationals of Poland and Yugoslavia accounted for 2 per cent respectively, nationals of EU countries for 6 per cent, of other European countries for 3 per cent and of non-European countries for 8 per cent, according to data published by the Employment Office.

5   The Northeastern Region is one of seven new regions within Hungary.

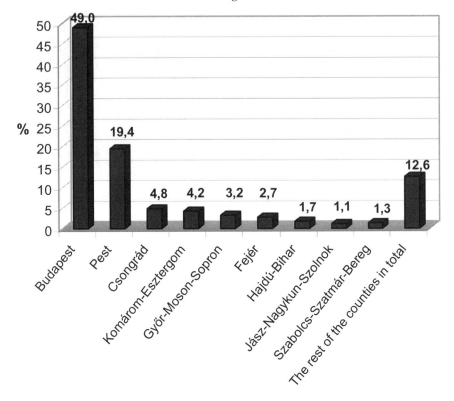

**Figure 12.2   The distribution of foreigners with valid work permits by counties
(31 December 1999)**

Only 530 foreign workers out of 41,972 with a valid work permit on 30 June 2002
were employed in the northeastern border counties of Szabolcs-Szatmár-Bereg.
The reasons for this low number were low wages even by Hungarian standards and
high unemployment levels, especially of unskilled labour, prevalent in the region.
The majority of foreign workers were skilled and worked as shoemakers or glass-
blowers (Romanians), needlewomen (Ukrainians), nurses or language teachers for
ethnic minorities (Romanians and Ukrainians); work permits for unskilled labourers
were issued very rarely.

Illegal employment, estimated to be much more important, is mainly limited
to seasonal agricultural and construction work. Small enterprises benefit from this
in order to recruit labour which would otherwise not be available and to cut labour
costs: locals will not always accept the jobs offered, and contractors for day-labourers
pay less for the accommodation of foreign workers than they would for the daily
transport of Hungarian labourers from outlying villages without access to public
transport. For many foreign workers, on the other hand, the bureaucratic procedure

Map 12.1   Communities studied during the survey

to obtain a work permit is too onerous.[6] Despite high levels of local unemployment, the illegal employment of foreign workers thus cannot simply be seen as detrimental to the local economy or to the employment prospects of the local unemployed. In addition, it appears that Hungary's eastern border area serves as a 'springboard' rather than being the final destination for job-seekers arriving from neighbouring countries (see Baranyi et al. 1999; Balcsók and Dancs 2001).[7]

According to survey data collected in four municipalities of Transcarpathia, as the Ukrainian part of the border region is called, 72.3 per cent of the inhabitants have some link with Hungary. The figures can be as high as 90 per cent in the case of Nagyapalád and Mezőkaszony, composed almost entirely of ethnic Hungarians, or 66 and 50 per cent in Csap and Tiszaújlak, where Ukrainians are well-represented. Conversely, only 22.1 per cent of ethnic Hungarians declared having no ties with Hungary, whereas this proportion reaches 71 per cent for ethnic Ukrainians. Only 5 per cent declared that they were not familiar with the closest municipality across the border, and 42 per cent that they were engaged in activities with neighbouring communities in Hungary: 28.4 per cent on a daily basis, 12.9 per cent weekly,

---

6   A work permit costs approximately 40,000 Hungarian forints, or € 160, with a waiting period of up to two months. Illegal workers are brought across the border in groups, and several hundreds of them have, for example, been observed to appear in the town of Fehérgyarmat to take up seasonal jobs.

7   Special mention must be made of questionnaire surveys conducted in 1999 and 2002 by the Centre for Regional Studies (Debrecen) in 119 municipalities of the border counties of Hajdú-Bihar and Szabolcs-Szatmár-Bereg as well as in 8 municipalities of the Hungarian-Ukrainian border area, half of the latter in Ukraine.

10.9 per cent monthly and 31.3 per cent only rarely. Among those with ties to Hungary, the reasons for crossing the border were shopping (59.5 per cent); visiting relations (53.1 per cent), friends and acquaintances (41.1 per cent); petty trade (35.9 per cent); employment (22.7 per cent, of which 53 per cent without a work permit); leisure activities (14.7 per cent); studies (8.3 per cent) and business relations (5.8 per cent).

Conclusion

Hostility towards 'foreigners' is usually explained as arising out of a sense of threat to one's livelihood and living standard and as being exacerbated by an economic context in which employment has become increasingly uncertain, precarious or rare. The underlying assumption is that foreigners are prepared to work for a fraction of the wages paid to citizens, thus being more competitive. This description fits Hungary's eastern regions where unemployment levels have been persistently higher than the national average since the early 1990s, and even more the border county of Szabolcs–Szatmár–Bereg (see Map 12.1), which is part of an economic peripheral region that extends into neighbouring Romania and Ukraine (see Baranyi in this book).

It should therefore not come as a surprise that the high numbers of registered and increasingly long-term unemployed have aroused local apprehensions towards job-seeking foreigners, although these fears are rarely based on personal experience but mostly on the representation of employment problems in the media.

However, as the above data show, the presence of foreign workers in most cases is not detrimental to the employment prospects of locals. First, overall numbers of foreign workers are low when compared to the capital and its surroundings or to other regions in western Europe; this can be attributed partly to Hungary's restrictive labour legislation. Secondly, those foreign workers who have obtained a work permit are in general skilled and not readily replaced by Hungarians. Thirdly, illegal employment appears to be mainly seasonal and confined to the agricultural and construction sectors which are less attractive to Hungarian citizens. Wages there are low, and the lack of adequate transport servicing isolated villages makes access to this labour market difficult for many locals. It is therefore doubtful whether repressing illegal employment of foreigners would lead to a decrease in local unemployment levels. To counterbalance biased information in the media, greater efforts should be made to inform citizens accurately about the role of foreign labour and 'illegal' labour migration for the economy.

Chapter 13

Local and Regional Cross-Border Cooperation between Poland and Ukraine

Katarzyna Krok and Maciej Smętkowski

This essay discusses cross-border cooperation between Poland and Ukraine within the new geopolitical situation in Europe. Polish membership in the European Union and transformation of the Polish-Ukrainian border into an external border of the EU have unambiguously influenced relations and cooperation patterns within the cross-border region. Fears of new exclusionary policies and a 'Fortress Europe' syndrome threaten to encumber the local cross-border relationships that have developed between the two countries since 1991. In order to provide a picture of regional cooperation under these new conditions it is therefore important to analyze both the scale and the scope of these changes and the new role of different actors involved in the cross-border cooperation.

The essay will provide an overview of previous cooperation patterns and experiences. It will also discuss future prospects of Polish-Ukrainian cross-border cooperation. In doing this, the socioeconomic and political conditions within which cooperation is emerging will be compared for both the Polish and Ukrainian border regions. Based on this regional overview, local dimensions of cross-border cooperation will be analyzed. The results of studies carried out in selected towns situated both on the Polish and Ukrainian side of the border will be presented and contrasted. This comparison allows the authors to provide deeper insights into practical aspects of different types of cross-border interaction and to identify the most important factors responsible for success or failure of public polices in this field.

Ukrainian Independence: The Starting Point

Although, in the past, the whole of Europe was characterized by constantly shifting borders (Blatter and Clement 2000; O'Dowd 2003), the Polish-Ukrainian border region was particularly affected by such changes, and for centuries its inhabitants have shared a rather chequered history. During the twentieth century alone there were five such changes. The last, in 1945, resulted in the entire region being broken up into two separate parts, one belonging to Poland and the other to the Soviet Union. This, in turn, led to the severing of existing social and economic ties. The administrative structure changed, and regional urban centres lost their spheres of influence. Thus,

Lviv, the largest urban centre of prewar southeastern Poland, became a Soviet city in 1945 and finally, in 1991, a major Ukrainian city. On the Polish side of the border, its former role as a main administrative centre was taken over by other cities, especially Rzeszów and Przemyśl. A formerly quite central area thus became provincial and peripheral. In addition, the new border was almost hermetically closed, and the lack of a good road and rail network limited contacts between those living on both sides of it. Indeed, after the second world war the main function of borders in central and eastern Europe was demarcation and defence. This was in stark contrast to western Europe where the progress of integration encouraged closer contacts and cross-border cooperation and where the function of borders became increasingly symbolic.[1]

When, after the collapse of the Soviet Union, Ukraine proclaimed its independence on 24 August 1991, Poland was the first country to recognize the new independent state. This event can be seen as the starting point for the development of cross-border cooperation between the two countries: the first bilateral agreement on mutual economic cooperation was signed the same year,[2] followed a year later by the Treaty on Good Neighbourliness and Friendly Relations, along with a number of agreements on cooperation intended to implement it. Relations at government level were excellent at the time. However, the lack of practical ideas for development as well as of experience in joint actions limited contacts to the highest level and conveyed to them a rather symbolic character. Furthermore, Polish foreign policy increasingly focused on cooperation with Western countries with the aim of the country joining NATO and the EU, while it was neglecting relations with neighbours to the east. Nonetheless, as a sign of goodwill in the area of regional cooperation, the Carpathian Euroregion, the first Euroregion outside the EU, was established in February 1993,[3] and a second, the Bug Euroregion, in 1995.[4]

---

1   Differences in the perception of the desirable character and functions of state borders which could be observed following the second world war still continue to influence the way people in western and eastern Europe see the role of borders. Studies thus show that people in western Europe place special emphasis on the social integration of border areas, the transfer of sovereignty and cross-border cooperation, whereas their counterparts in eastern Europe are likely to focus on state security and sovereignty as well as processes of inclusion and exclusion (Berg 2001).

2   The Agreement between the Government of the Republic of Poland and the Government of the Republic of Ukraine on trade and economic cooperation was signed on 4 October 1991 and came into force on 11 March 1993.

3   The Carpathian Euroregion was established by the ministers for Foreign Affairs of Poland, Ukraine, Hungary and Slovakia, and was enlarged in 1997 to include several border areas in Romania.

4   A first attempt to establish a Euroregion was made through the Regional Agreement between the Bialskopodlaskie, Chełmskie, Lubelskie, Tarnobrzeskie and Zamojskie voivodships in Poland, the Bresk *voblast* in Belarus, and the Volyn *oblast* in Ukraine, signed on 4 July 1992. The Euroregion was finally established in September 1995 through an agreement between the above-mentioned parties, except for the Bialskopodlaskie voivodship and the Bresk *voblast*, both of which joined at a later date, in 1998. In 2000 the Bug Euroregion was further enlarged to include the Sokoliv and Zhovkva *rayons* of Lviv *oblast* in Ukraine.

The early removal of visa requirements was clearly contributing to the development of mutual cooperation and to the promotion of contacts between various Polish and Ukrainian partners at regional and local levels. The number of people crossing the border increased dramatically, mostly because Ukrainians wanted to take advantage of different price levels.[5] This regime was, however, abolished on 1 October 2003 when Poland, then a EU candidate, was obliged to introduce visas for citizens of neighbouring non-EU countries.[6] So far, Ukraine has been Poland's only post-Soviet neighbour to agree to non-reciprocal visa arrangements whereby Poles are able to enter Ukraine without a visa while Ukrainians need one to enter Poland, although these visas are granted free of charge and the application procedure has been reduced to a bare minimum.

Structural and Administrative Aspects of the Border Region

The 529-km long Polish-Ukrainian border is Poland's third-longest and Ukraine's fifth-longest state border. The border region, as discussed here, is comprised of four major administrative entities (or provinces): the Lubelskie and Podkarpackie voivodships on the Polish side, and the Volyn and Lviv *oblasts* on the Ukrainian side. It has a surface of 85,018 km^2 and a population of 8,060,300. Population density is low and the region is little urbanized: only 43.9 per cent of the inhabitants on the Polish side and 56.9 per cent on the Ukrainian side live in towns. A fall in the natural birth rate, still more pronounced in urban areas, – and a higher mortality rate in Ukraine – is one of the major reasons why the population has been stagnating between 1990 and 2001.

Border areas on both sides have a significantly lower share in the gross domestic product of their country and have been developing at a slower rate. Gross added value, for instance, is lower than the national average – by 30 per cent in Poland and by 20 per cent in Ukraine. About 40 per cent of the active population are working in the agricultural sector, which is composed of small farms that are neither market-oriented nor profitable. Rising unemployment levels during the second half of the 1990s, which have reached an unprecedented 17 per cent in recent years, have led to a high negative migratory balance on the Polish side, and there are comparatively fewer businesses than elsewhere in Poland, despite an increase in absolute numbers. In Ukraine the collapse of the Soviet Union in the early 1990s resulted in a dramatic fall of living standards; the gross domestic product dropped by 58 per cent between 1990 and 1995, pushing people to leave their region or even the country. Although the unemployment rate has been reduced to some 4.5 to 7 per cent in the recent

---

5 Considerable economic disparities, as defined by the 'capacity flow' classification applied in the ESPON 1.1.3. project, encourage cross-border activities at grassroot level (see European Spatial Planning Observation Network 2004).

6 New member states were supposed to form a sort of buffer zone between the old member states and non-EU countries, a type of border regime referred to in the literature as 'the march' (see Foucher 1999; Kramsch and Hooper 2004).

Table 13.1   Social and territorial structure of the border region (2001)

	Poland (Lubelskie and Podkarpackie)	Ukraine (city of Lviv, Lviv and Volyn region)
Area (in km2)	43,041	41,977
Share in national territory (%)	13.8	7.0
Population	4,358,900	3,701,400
Share in national population (%)	11.3	7.6
Population density (inhab. per km2)	101	88
Urbanization index (%)	43.9	56.9

Source: *Statistycznyj szczoricznik Ukraine za 2001 rik* (Kiev, 2002) and *BDL - bank danych lokalnych* (GUS 2001).

past, this has not meant higher employment levels. Many of the unemployed simply appear not to register any more or to have emigrated. However, new businesses have been created at the stable national rate of 7 per cent.

Marked differences in territorial administration between the two countries significantly affect cross-border activities. Ukraine is a more centralized country than Poland, and local governments there are handicapped by their limited competence and scope of activities. Finances are under strong central control barring local authorities from gaining more independence. More generally, centralized decision-marking has led to protracted administrative procedures, and the state authorities sometimes consider as irrelevant, or even contradictory to their own policies, local and regional development goals and activities.

In order to determine the impact of cross-border cooperation, we have analyzed official data on border traffic and foreign trade and conducted field research in seven selected towns of the border area: Przemyśl (68,000 inhabitants) and Jarosław (40,000), Zamość (67,000) and Tomaszów Lubelski (20,000), in Poland, and Yavoriv, Zhovkva and Sokal, three smaller towns in western Ukraine, which is dominated by the metropolis of Lviv.[7]

Przemyśl and Zamość were provincial capitals until the administrative reform of 1998, when their functions were taken over by Rzeszów and Lublin. Their economy

---

7   Research was conducted within the framework of the EXLINEA programme by a team composed of G. Gorzelak, B. Jałowiecki, E. Kozłowska, K. Krok, A. Olechnicka, O. Mrinska and M. Smętkowski. The survey in Przemyśl is a follow-up study to research conducted in 2001 by a team composed of G. Gorzelak, M. Smętkowski and A. Tucholska and included a questionnaire submitted to representatives of more than 200 businesses in Poland and of some 40 businesses in Ukraine; data collected at stalls in open-air markets and accommodation facilities; in-depth interviews and various other questionnaires.

Black = towns selected for field research

Map 13.1   The Polish-Ukrainian border regions

is based on transport, trade and non-market services, especially tourism, while manufacturing and construction are less developed. Tomaszów Lubelski, the smallest of the surveyed towns, is located along main roads leading to the three main crossing points of this section of the Polish border at Medyka, Korczowa and Hrebenne and is the seat of the county *(poviat)* authorities, which accounts for the high share (almost 50 per cent) of non-market services in its economy. It also has a fairly large number of businesses operating in the manufacturing and construction sectors. Jarosław, on the other hand, is an industrial centre of more than regional importance. Its huge plants have been privatized, partly with foreign capital, and the town has a higher than average number of small and medium-sized enterprises owned by physical persons in comparison to other towns of the Podkarpackie region.

Entrepreneurs in the surveyed Polish towns generally prefer trade to manufacturing: in Przemyśl, and to a lesser extent in Jarosław, the number of businesses registered with the national economic registry (REGON) is considerably lower than the national average in the manufacturing and construction sectors and higher for sectors such as trade, repairs, hotels and restaurants, transport, storage and communications, other things being equal. In addition, most enterprises in the manufacturing sector produce for the local market, while only a few are looking for partners in western Europe and practically none east of the border. Goods originating in Ukraine represent only 1 per cent of total supplies for Polish enterprises. Few enterprises declare importing low-processed goods or raw materials from Ukraine, and more attention is being paid to identify export markets in western Europe, which receive less than 20 per cent of the production at present. The strategy of traders who sell clothes, footwear, furniture and building materials brought in from central Poland to Ukrainian customers consists more in flexible adaptation to demand than possible investments in manufacturing these goods. In all surveyed towns the business environment is poorly developed: financial intermediaries, real estate and other business-related activities are less present than elsewhere in Poland; only Jarosław is actively promoting and implementing measures to attract potential international investors and partners.

The three Ukrainian towns and their counties *(rayons)* all belong to different types of the country's industrial zones: predominantly manufacturing (machinery, food and light industry) and construction for Sokal; mostly chemical, woodworking, pulp and paper industries as well as manufacturing of simple machinery for Yavoriv; and light, food and woodworking industries as well as manufacturing of construction materials and simple machinery for the least industrialized and least diversified economy of Zhovkva.

Cross-Border Traffic

The Polish-Ukrainian border, and Poland's eastern borders in general, do not have sufficient facilities to effectively service border traffic. Of the 16 existing crossing points linking Poland to Ukraine (ten by road and six by railway), two – Dołhobyczów-Novoukrainka and Malhowice-Nizhankovichi – handle only simplified and much reduced border traffic, and passenger traffic by train has fallen from an initial 42 per cent to a low 4.5 per cent in 2003.[8] Long waiting times at crossing points – currently up to 3 or 4 hours – are a major barrier to the free movement of persons.

Fluctuations in border traffic since the opening of the border in 1991 have been due mainly to macroeconomic changes in both countries as well as to the repeated introduction of new regulations tightening or loosening entrance requirements for persons or goods (see Figure 13.1). Thus, records for the first year show some

---

8   The most important border crossing by road is at Medyka, near Przemyśl. Until 1991, Medyka was practically the only border crossing by road between Poland and Ukraine (then still part of the Soviet Union). The opening of a new crossing point at Korczowa in 1997, intended to reduce traffic at Medyka, led in fact only to a loss of traffic at the more northerly crossing points of Dorohusk and Hrebenne.

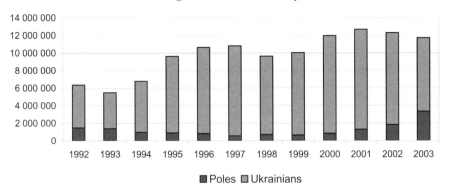

Figure 13.1   Polish-Ukrainian border traffic, 1992–2003

7.4 million crossings, mostly by Ukrainians. Significant differences in the supply and price of several commodities rapidly led to the creation of numerous open-air markets on the Polish side, where Ukrainians sold goods that were lower taxed in their own country, such as cigarettes, alcoholic beverages and petrol, and purchased – mainly manufactured – goods that could not be obtained at home, such as chemicals, cosmetics, clothes and furniture. Petty trade became the major form of cross-border cooperation. As a corollary, there was also a rise in the number of people employed to provide services for these traders. However, this did not result in any sustained development.

In 1998, the open-air market at Przemyśl, for instance, was one of Poland's largest. But since then, the number of stalls, though still important, has been more than halved (from 1,500 in 1997 to 700 in 2003), and the main customers are now Poles.[9] As a consequence, the municipality has suffered a considerable loss in revenue from fees: in 1997, these accounted for PLN 4.1 million (or 7.3 per cent of its budget), and represented only PLN 0.7 million (or 1.3 per cent) in 2002. Moreover, much of the trade has always gone unregistered or been illegal. Similar though less drastic changes can be observed for markets in the other towns surveyed. Most of them have become less competitive as the former petty trade has been taken over by Ukrainian whole-sellers, and goods such as cigarettes and alcohol are now often bought by Poles in Ukrainian markets.

One would expect a location in the border area to be a stimulus for developing services offered to travellers, such as shops, bars, restaurants, hotels, filling stations, currency exchange outlets or passenger transport, though much would depend on the purchasing power of the customers and the purpose of their journey. In all cases studied, travellers crossing the border area in transit show little interest in the

---

9   Whereas in 2001 Ukrainians were declared to be the main customers of 24 per cent and the second most important ones of 65 per cent of the stalls, these numbers have dropped to 15 and 50 per cent respectively in 2004.

available offer. The share of foreigners in the volume of turnover created by trade outlets and eateries is small (Smętkowski 2001), and hotels mainly serve Polish guests. In Przemyśl, for instance, Ukrainians accounted for only 25 per cent of the number of hotel guests and for just 15 per cent of turnover in 2001 (ibid.); in 2004, their share did not exceed 12 per cent, except in Tomaszów Lubelski (20 per cent). Even in Zamość, where the quality of accommodation has been rapidly improving owing to the city's attractiveness for tourists, Ukrainians account for only 5 per cent of hotel guests. Similarly Polish travellers seldom spend the night at hotels in Ukraine, except those located in the regional capitals, Lviv and Lutsk, and in some popular tourist destinations.

Services for traders in the Polish border region thus proved a short-lived phenomenon, strongly dependent on general economic performance and customs regulations. The number of border crossings fell repeatedly, for instance, in 1992 and 1993, after first economic reforms in Ukraine made part of the trade unprofitable, and again in 1998, in the aftermath of the Russian crisis, when Ukraine introduced stricter export regulations for certain goods, and as a consequence of Poland adopting a tighter border regime. Overall, the number of persons crossing the border has, however, been increasing, with a record high of 12.7 million in 2001 and 11.7 million in 2003. But whereas Ukrainians accounted for up to 95 per cent of crossings in the mid–1990s, their share has dropped to 71.5 per cent in 2003. With the decline in petty trade the share of Ukrainians living outside the border region fell from 55 per cent in 1997 to 31 per cent in 2002.[10] Moreover, an increasing number of Ukrainians now cross the border in transit for western Europe. Meanwhile, greater numbers of Polish tourists, slightly more often from outside the border region than before, are visiting places of historical interest or are making use of cheaper recreation and rehabilitation facilities in Ukrainian spas such as Truskawiec. Furthermore, since 2000, Poland has become a key foreign investor in Ukraine after accelerated economic growth and the implementation of various reforms improved the local business climate and, more generally, the country's image.[11] More recently, in October 2003, Poland's accession to the European Union led to a sharp drop of crossings by Ukrainians but does not appear to have affected long-term trends.

Foreign Trade and Cross-Border Trade

With a turnover of US\$ 2.3 billion in 2003, Ukraine is Poland's fifteenth-largest trade partner and her second-largest in eastern Europe, after Russia. More importantly, Ukraine is one of the few countries with which Poland has a clearly positive balance of trade in commodities (US\$ 0.8 billion in 2003). Polish exports to Ukraine started

---

10 The number of Ukrainians spending the night in Poland has decreased by half; the fall was particularly sharp in the northern part of the border region.

11 By 2004, 339 Polish companies had invested some US\$ 60 million (or 18.6 per cent of all foreign capital investment) in the Lviv *oblast*. The vast majority of these investments were made after 2000.

to take off in 1994 to reach US$ 1.2 billion worth of goods in 1997, fell by 35 per cent to US$ 0.7 billion during the economic crisis in Russia and have since been increasing steadily. Ukrainian exports to Poland, on the other hand, have known a more stable, but much less impressive growth rate – from some US$ 0.2 billion to US$ 0.45 billion worth of goods – between 1994 and 2002; however, by 2003, they have risen to US$ 0.75 billion, probably as a result of recent foreign investment in western Ukraine (see Figure 13.2).

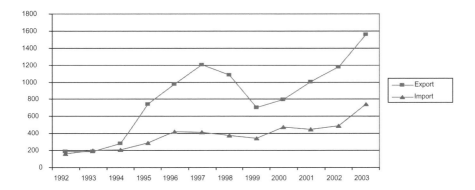

Figure 13.2   Value of Polish-Ukrainian foreign trade, 1992–2003 (in million US$)

Polish exports to Ukraine include mainly products from the electrical engineering (33.7 per cent) and chemical (21.4 per cent) industries, whereas imports from Ukraine are even less varied: metallurgical products (41.1 per cent) and fuel and energy (27.6 per cent) predominate. This trade pattern indicates that foreign trade between the two countries is based on industrial centres in central Poland and eastern Ukraine.

When compared to the national level, data on exports from the Polish border region show a higher share of little-processed goods from the light industry (Podkarpackie) and of processed food as well as of fuel and energy (Lubelskie), while main imports include agricultural and forest produce as well as chemicals (Podkarpackie) and foodstuff (Lubelskie), in addition to fuel and energy. After a drop of 12.5 per cent between 2000 and 2003, the border voivodships' share in the exports (see Fig. 13.4) now represents about a quarter of the value of Poland's exports. As the region contributes only 8 per cent to the gross domestic product, it is likely that companies from the border region regularly serve as intermediaries in the trade between central Poland and Ukraine, a hypothesis which has been corroborated by findings from our survey (see Table 13.2). Only one-tenths of imports of Ukraine end up in the border region.

A comparison between the markets for supplies and sales in both countries confirms this asymmetrical relationship: Ukrainian enterprises operating in the border

Table 13.2   Trade turnover of Polish border voivodships with Ukraine

Lubelskie

	2000	2003	Trend
Exports (in million US$)	136	145	106.5
Share in national exports (%)	17.0	9.3	—
Imports (in million US$)	32	43	134.7
Share in national imports (%)	6.7	5.8	—

Podkarpackie

	2000	2003	trend (%)
Exports (in million US$)	160	238	148.9
Share in national exports (%)	20.1	15.3	—
Imports (in million US$)	9	28	298.7
Share in national imports (%)	2.0	3.8	—

Note: Between 2000 and 2003 Poland's exports increased from US$ 798 million to 1,591 million (increasing to 195.6 per cent) and its imports from US$ 475 million to 745 million (156.6 per cent).

Source: prepared by the authors based on data from the Foreign Trade Data Centre (CIHZ) prepared by T. Komornicki.

region bought the bulk of their supplies from Polish partners and sold their produce in other than Polish markets, whereas Polish companies rely less on foreign supplies and exported their produce mainly to Ukraine. The diminishing significance of this trade pattern is probably due to a 'squeezing effect' (Smętkowski 2002), whereby Polish enterprises loose their competitive advantage once production facilities with more modern technology are set up in Ukraine, which has lower labour costs than central Poland.

Polish-Ukrainian Cooperation – the Local Actors' Perspective

It was local government institutions on the Polish side who first took the initiative to establish contacts with partners on the other side of the border. Initially these contacts were of a symbolic nature – paying courtesy visits or signing bilateral partnership agreements – and some have remained so until today. Nonetheless, they facilitated cooperation developed by other local organizations and actors. Polish local officials today consider private entrepreneurs to be the most active partners in cross-border cooperation. The same cannot be said with regard to Ukraine where the major actors are the state regional administration and the regional council; it is they who have a genuine impact on decisions leading to the planning and implementation of cross-border activities. Local authorities declare to have little say in such matters, even though they are most interested in establishing cooperation and achieving tangible results. It must also be stressed that activities within the framework of Euroregions, which by definition should actively promote and develop cross-border cooperation, do not have a very broad regional audience. This is particularly true for the Carpathian Euroregion of which very many people have never heard (see Béla Baranyi in this book). Activities of the second, more northerly Euroregion Bug are viewed more favourably. Such critical opinions can be explained by the little support and sparse funds these Euroregions received from the Union.

Indeed, cooperation between local governments and institutions have to date been on a small scale only. Projects are of a 'soft' nature and included activities such as the organization of meetings, training, peer learning, consulting and missions. Furthermore, their effectiveness is judged low by those who are supposed to benefit from them, with some variation according to the region. In the northern part of the Polish border region, the former Lublin voivodship, local representatives have been unable to mention a single completed project or claimed that no joint activities other than planning has taken place, due to a shortage of funds or programmes which could have provided a framework for them. In the Podkarpackie voivodship and on the Ukrainian side of the border, representatives listed several successful joint activities in the fields of tourism, culture and environmental protection such as the Galicia Festival on the Polish side, the Roztocze Festival in Zhovkva, joint development of bicycle paths, and ecological development and recultivation of the 'Sirka' sulphur basin in Ukraine. However, it must be stressed that public authorities and NGOs involved in the promotion of cross-border cooperation are often insufficiently informed about actual interactions and relationships.

The situation has slightly changed since 1998, when the PHARE Small Projects Fund (SPF) was being launched with the aim of creating possibilities for co-financing initiatives by Euroregions. However, only a dozen or so projects have so far been financed in this way, and EU funds allocated to this border area have remained quite modest (Krok 2004).

Generally speaking, EU assistance to Polish border regions before the country's accession mainly benefited its western regions. During the initial budget years (1994–99) the PHARE-CBC programme for the Polish-German border regions made use of some € 272 million, while the entire eastern border area received only € 48 million during the initial budget years (1996–99) of the PHARE-CBC Integrated Eastern Border Programme, most of which was spent on making the border safer and less permeable.[12]

When asked about cross-border cooperation, both local officials and entrepreneurs agreed that regulation and the improvement of border crossings should receive a high priority. One issue which was unfavourably assessed was the obligation for

---

12 Based on data provided by the PHARE CBC Implementing Authority.

Ukrainian citizens to carry a minimal amount of money upon entering Poland, something considered a more severe hardship than the requirement of a visa. Other favoured issues revolve around cooperation to improve infrastructure for tourism (extension of existing bicycle paths, common tourist trails, creation of cross-border nature reserves, etc.).

Poles and Ukrainians differ, however, with regard to the final aims of cross-border cooperation. Ukrainian officials regard it as a process intended to familiarize citizens with the idea of a 'shared Europe', EU rules and mechanisms, and to create more awareness of opportunities offered by European integration. Emphasis, apart from purely economic aspects, is being put on the transformation of the current meaning of the border from being a dividing line ensuring the state's sovereignty into a link connecting areas on both sides of it. Polish representatives do not consider this particular issue as being of great importance and are looking at cross-border cooperation in more practical terms.

Among the few measures fostering Polish-Ukrainian cooperation the most prominent ones are fairs and exhibitions. Although these are popular with entrepreneurs, who complain that there are not enough of them, they are – paradoxically – judged by them as being not particularly useful. Training schemes for cross-border cooperation are similarly rated and ranked second, while information about the conditions of cross-border cooperation, ranking third, is considered to be insufficient by Polish entrepreneurs, but favourably assessed by Ukrainians, who declare that they frequently benefit from informal assistance by the authorities in obtaining privileges for border crossing. Economic missions are thought to be well-established instruments of cooperation but, again, appear to be of very little use to entrepreneurs. Conversely loan guarantees were viewed very positively, despite their limited availability.

Besides various obstacles linked to individual and passenger border traffic, entrepreneurs list corruption, public insecurity, bureaucracy and uncertain regulations as the major barriers to trade. In Ukraine, bureaucracy and corruption, though less critically judged by Ukrainians than by Poles, are considered the main problems. The issue of public safety receives a similar rating by Polish and Ukrainian entrepreneurs, but the latter view it less critically and express only moderate criticism of bureaucracy, uncertain regulations and corruption on the Polish side of the border, tending to play them down in the same way they are underestimating various obstacles of a financial nature. They also complain about insufficient transport infrastructure in Ukraine, particularly when compared to Poland. All entrepreneurs declare that they seldom have recourse to instruments fostering trade cooperation between Poland and Ukraine. But it remains unclear whether these instruments are missing because of the insignificant scale of trade or whether trade has not developed because public authorities did not implement such instruments.

Polish and Ukrainian entrepreneurs also differ when asked to assess future prospects of cross-border cooperation. Ukrainian entrepreneurs thus have a more favourable opinion about Poland's accession to the EU, probably because they are hoping to benefit to some extent from easier trade contacts with other EU countries once the EU border has been brought closer to Ukraine, but also because they do not appear to share the concern of Polish entrepreneurs that EU accession could be an economic threat. Consequently, more of them expect the situation of their enterprises to improve and Polish-Ukrainian trade contacts to intensify. Polish entrepreneurs more frequently stress the impact of obstacles to border crossing, such as the visa requirements introduced in 1998 and in October 2003, possibly because these restrictions are creating barriers for the customers of the local retail and wholesale trade, whereas Ukrainian enterprises have not been directly affected by these measures. It is even possible that Ukrainian companies see in them an opportunity to improve their competitive position. Finally, Ukrainian entrepreneurs tend so see foreign trade with Russia and Poland as a sort of zero-sum game, where the Russian crisis has fostered the development of Polish-Ukrainian ties.

In the opinion of both Polish and Ukrainian entrepreneurs, Poland's accession to the EU primarily ought to curb corruption, reinforce the rule of law and improve public safety. It is also expected to improve transport, telecommunications and business infrastructure. Some Polish entrepreneurs fear that bureaucracy will increase, while their Ukrainian counterparts hold the contrary view. However, it should be emphasized that most respondents mentioned neutral consequences or did not have any view on this question.

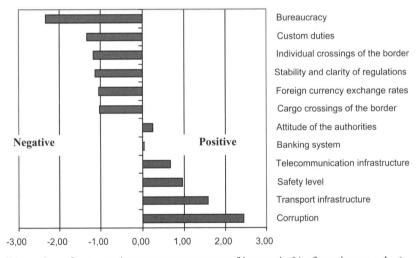

(% number of answers * average assessment of impact in % of maximum value)

Figure 13.3   Impact of Poland's EU accession on the conditions of trade exchange with Ukraine as seen by Polish entrepreneurs

Conclusion

Neither the region nor the surveyed towns appear to have gained from their location near the border or from the existing border infrastructure. Travellers in transit, for instance, rarely use local accommodation and catering facilities. Only transport services benefit to a certain extent from day trips organized for trade purposes. Polish tourists visiting Ukraine, though on the increase, have no real impact on the local economies outside that of Lviv, their main destination and the region's largest urban centre.

Despite close cultural and historical ties, Poland and Ukraine are today wrestling with a broad array of problems. It seems evident that Ukraine still has to address many systemic and social issues Poland already solved in the recent past. It might therefore be that Ukraine is now following a path covered by Poland in the 1990s. Indeed, there are many similarities, in particular with regard to socioeconomic and administrative factors, that were present in cross-border cooperation on the Polish-German border during the 1990s.[13]

There exists a significant difference in the level of development between Poland and Ukraine. Poland's eastern neighbour has a lower GDP per capita, lower labour costs and mostly lower prices. Its market is not yet saturated even with basic goods and services, which leaves room for the development of cross-border trade but does not ensure its intensification. Poland's border region is economically undeveloped and therefore unable to produce innovative goods and services for the international market, including Ukraine. For this reason cross-border cooperation so far has mainly relied on trade brokerage between manufacturers from central Poland and Ukrainian purchasers, creating a highly asymmetric trade pattern with great surpluses for Poland selling processed goods and buying simple products and raw materials.

Important price differences for a great number of products propelled the initial development of petty trade in the border region, taking the form of open-air markets. Limited purchasing power of the region's population as well as the gradual levelling out of prices led to a decline in the profitability of this type of trade, which is now restricted to a small number of (often smuggled) goods, such as alcohol, cigarettes and petrol. Moreover, after 1998, the negative economic impact of the Russian crises, on the one hand, and new restrictions imposed on border traffic, on the other, contributed to further decline. Currently, it can be observed that this highly volatile form of trade is being replaced by official, registered trade with an increasing turnover, as can be deduced from the increasing number of lorries crossing the border. This could prove to be the first stage of a process leading to enduring economic and social ties as well as to the development of lasting cross-border cooperation (Perkmann and Sum 2002). Although Poland's accession to the European Union so far has not affected economic exchange in any significant way, EU enlargement is commonly perceived as a stabilizing factor with a favourable influence on trade between the two countries. The influx of EU funds and the adoption of institutional arrangements

---

13 See Scott and Collins 1997; Krätke 2002; Gorzelak, Bachtler and Kasprzyk 2005.

linked to them, including a boost to the Euroregions' activities, might help reduce existing obstacles to cross-border cooperation.

Although the experiences of the last years demonstrate that cross-border cooperation has been sluggish and rather disappointing in terms of positive impacts on regional development, events since 2004 show promise. Ukraine's agenda of democratic reform and renewal as well as closer and improved ties to the EU will provide a greatly improved political environment for cross-border cooperation. As Ukraine narrows development and 'governance' gaps with Poland, structural and administrative conditions for more successful cooperation will also improve. Of course, cross-border cooperation is by no means a panacea for entrenched regional development problems. However, it can help bring about a much-needed normalization and enhancement of political dialogue and general social exchange between both countries.

PART V
Cross-Border Cooperation and Regional Development at the Former External Borders

Chapter 14

Normalizing Polish-German Relations: Cross-Border Cooperation in Regional Development

Grzegorz Gorzelak

The present-day German-Polish border is of recent creation. It was established in the wake of the second world war by the victorious allied powers and brought about the forced westward expulsion of millions of Germans and Poles. Suffice it to say that a genuine German-Polish rapprochement and a political normalization of the border were only possible after the momentous political changes of 1989–1990 and the conclusion of new bilateral treaties reaffirming the will of both countries to promote peaceful coexistence. The German-Polish Treaties of November 1990 and June 1991 reconfirmed the Oder and Neisse Rivers as the permanent boundaries between the two countries and established cooperation frameworks based on principles of good neighbourliness and peaceful coexistence. They represented essential steps for a genuine German-Polish rapprochement and an indispensable precondition for the development of cross-border planning cooperation. This fact alone indicates that the construction of a German-Polish cross-border region – irrespective of the spatial level involved – has remained, and for some time will remain, a highly artificial and yet necessary and desirable project. While there can be no doubt that economic problems, structural deficiencies and prejudices continue to represent major challenges for cross-border cooperation, I argue that EU policies have promoted learning processes, such as those advocated by Hans-Joachim Bürkner in this book. In other words, I contend that there is a need for a sense of belonging to a common neighbourhood and region, born of Polish and German history. I also contend that, despite the existence of 'exclusionary' mentalities within the region, impacts of all policies and institutions can be identified that have been at work since the opening of the border in 1991. Cross-border cooperation has offered considerable potential for developing local government capacities, local networks and civil society initiatives.

In order to fully understand present-day contexts of German-Polish cooperation, it will be necessary to discuss certain historical and geopolitical issues, starting with the complex territorial shifts the Polish state and its boundaries have undergone over the centuries. The essay then focuses on the management of EU cooperation initiatives, notably INTERREG and PHARE-CBC, and on the effects of these programmes on local governance. As I argue, western Poland was integrated into the EU long before Poland's formal accession in 2004 through collaboration with German counterparts. The next section then discusses attitudes of the local citizenry towards their neighbours and concludes that tenacious stereotypes are slowly giving way to a more 'open' notion of region. Finally, I venture some conclusions about German-Polish cooperation as a possible general 'model' for cross-border cooperation, particularly in view of other, more difficult contexts on the EU's new external borders.

The Shifting Geography of the German-Polish Border

The Polish-German border has been one of the least stable in all of Europe. But the original border that separated Slavonic and German peoples more than ten centuries ago is in fact not all that different from the present Polish-German border. Slavonic tribes lived on the southern coast of the Baltic Sea, with settlements extending to the Oder River, and even as far as and beyond the Elbe River, in the west. After repeated attacks from Charlemagne and, later, from German kings and princes, the Oder River became the effective border, much in the same way as it separates Germany and Poland today.

The series of maps reproduced below detail the evolution of the Polish-Prussian border until the late eighteenth century. They show a rather dynamic situation brought about by the rise and fall of the Polish kingdom and by geopolitical conflicts between Austria, Russia, Sweden and Prussia.

Even more dynamic were the Polish-Prussian and, after 1870, Polish-German border relations. Changes made to state boundaries during the period of Poland's three partitions (1772, 1792 and 1795) and in the recent past are shown in Map 14.1.

Of most general importance to the present discussion are the changes that were made to the Polish-German state border after the Treaty of Potsdam, in 1945. Five crucial issues can be identified in this regard.

1. An almost complete transfer and change of population took place in the territories denoted as '3' in Figure 2. On the basis of the Potsdam Treaty, the German population was either asked to leave or forcibly expelled. There are no official estimates of the magnitude of this exodus, but several million persons were involved. The German population was replaced by Poles who either were themselves expelled from eastern areas of pre-war Poland lost to the Soviet Union or voluntarily migrated from other parts of Poland.
2. Until 1989 the Polish-German border was an international boundary between two states that belonged to the Soviet Union's sphere of influence. This 'border of friendship' was in fact mostly a border of separation, allowing only sporadic interaction between communities on both sides of it.
3. The Polish-German border, although widely recognized internationally, was officially acknowledged by the Federal Republic of Germany only on

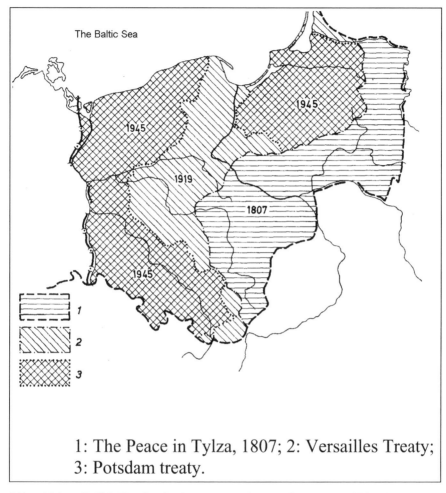

The Baltic Sea

1945

1945

1919

1807

1945

1

2

3

1: The Peace in Tylza, 1807; 2: Versailles Treaty;
3: Potsdam treaty.

Map 14.1   Polish Territories lost to Prussia and Germany, 1803–1945
Source: Piskozub, 1987, p. 168.

7 December 1970 when Chancellor Willy Brandt signed a border treaty with
Poland. The validity of this treaty was subsequently confirmed during the
Paris Conference of July 1990 and specified in several agreements. A final
Polish-German treaty was signed on 17 July 1991.

4. The long period of legal uncertainty regarding the status of the German-
Polish border led to a deep feeling of insecurity among the inhabitants of
Poland's western and northern territories. It was widely supposed that
Poland's 'regained territories' might at some point be returned to Germany.
Therefore, the inhabitants of these territories refrained from making any major

investments or improvements on properties they occupied. This also led to an overly sensitive attitude toward German influence and presence in these areas. The prospect of Germans purchasing land and property in western Poland or receiving compensation for expropriated property was thus extremely unpopular.

5. Although the territories obtained from Germany after the second world war benefited from better infrastructure and quality of housing than the rest of Poland, the resettled Polish population came from regions with much lower material living standards, a situation which led to a process of 'mutual adaptation' during the first years after the war. Moreover, these territories were considered by national authorities as richer than the rest of the country and thus not in need of major investments. As a result, after three decades of neglect, material assets were highly depreciated. This remains a major obstacle to development in many parts of western Poland even today.

6. Until recently it was generally held that the inhabitants of Poland's 'regained territories' had not been able to create solid community ties nor to have established a deep-seated rapport with their natural environment. Sociological research does not confirm these hypotheses, and some findings (Gorzelak et al. 1999) indicate on the contrary that the communities of Poland's northern and western territories have very active local governments, although they suffer from a proliferation of social problems stemming from high unemployment, especially among the young, and are hotspots for illegal cross-border activities.

History thus provided very few favourable conditions for the development of Polish-German cross-border cooperation. In order to achieve a breakthrough many obstacles inherited from the past had to be overcome.

The Context of Postsocialist Transformation

After the collapse of the communist system in central and eastern Europe, the regions on both sides of the Polish-German border faced both serious challenges and budding prospects. The greatest challenge were the necessary social and economic transformations, and the two regional systems – Polish and German – passed through them in dramatically different ways and with contrasting results. Authorities on both sides of the border immediately came to the conclusion that close cooperation would benefit both sides. Programmes of joint action and planning emerged as early as 1991, and the famous 'Stolpe plan' proposed a complex agenda for the future development of the border region. However, these efforts proved somewhat premature as Polish authorities needed more time to accept a more open approach to Polish-German cross-border cooperation (Guz-Vetter 2002).

The major difference between the German and Polish border regions is the same as that between East Germany and Poland as a whole: the former has received large external subsidies (more than € 1 trillion, though obviously less in net terms), while the latter has had to struggle with the costs of economic and social restructuring on its own by turning to the open market for competitive advantages. Following the pattern of Schumpeterian 'creative destruction', Poland was the first postsocialist country to enter recession and, in mid–1992, the first to show the ability to achieve positive growth. External assistance was very limited and did not exceed € 1 billion at the moment of Poland's accession to the EU in 2004. If Poland were to have been subsidized at the same magnitude as eastern Germany, it would have received the equivalent of its yearly gross domestic product throughout the entire period since 1990. In spite of assistance received, the German border regions represent one of the more sombre cases in contemporary Europe – the economy is stagnant, unemployment high despite large-scale migration to the western part of the country and entrepreneurial spirit low.

On the Polish side of the border, we find a somewhat contrasting picture. The area is still stricken with high unemployment but maintains a certain economic dynamism. Regions in western Poland have had a mixed economic fortune. The collapse of the state farms, established after the war on former German latifundia, deprived these regions of one of their key economic sectors. In addition, employment opportunities for former workers of these state farms are very limited; job seekers have few skills and little motivation, and often live in peripheral locations. At the same time these regions have fared better than the rest of the country thanks to better infrastructure and their proximity to Germany. The latter was indeed one of the positive factors at work in the postsocialist transformation of Poland's western regions. Relatively poor East Germans, that is poor in comparison to their western compatriots, have started to cross the now open and friendly border in search of cheaper commodities and basic services. In the mid–1990s, their purchases amounted to about DM 5 billion (€ 2,5 billion), an important economic contribution to the economy of the Polish border region, its businesses, local governments and people, even though consumer demand decreased during later years – due to the appreciation of the Polish currency (Balassa-Samuelson effect) – before reaching once more higher levels. Furthermore, institutional experience brought to East Germany from the western part of Germany and later transferred to Poland within the framework of cooperation projects added to overall positive change in the western Polish regions. Early involvement of Poland's western regions in the EU-financed PHARE programmes, and especially the Polish-German CBC programme, was indisputably a positive factor in the postsocialist transformation and development of these regions.

The Polish-German Crossborder Cooperation Programme

The legal basis for the PHARE-CBC was the European Commission's Regulation No. 1628/94 of 6 July 1994 concerning the implementation of a programme for cross-border cooperation between countries in central and eastern Europe and member states of the Community. The Polish-German border territory eligible under the programme was defined in the Financial Memoranda for 1994-1999 and the Multiannual Indicative Programme. At the outset, this territory included the then voivodships adjacent to the German border, that is Szczecin, Gorzów Wielkopolski, Zielona Góra and Jelenia Góra. In 1995, the eligible territory was extended to the (then) voivodships of Koszalin, Piła, Poznań, Leszno, Legnica and Wałbrzych. The total territory of these voivodships covered over 68,900 km^2 (or 22 per cent of the entire territory of Poland) and comprised 451 municipalities (see Map 14.2).

In the period between 1994 and 2006, total resources allocated to the Polish-German CBC programme amounted to about € 400 million, of which € 280 million were granted during the first programming period which ended in 1999. This first period is of special interest, as it represented a first step in formalized Polish-German cooperation within the framework of the EU. The total outlay during this period was equivalent to only 0.19 per cent of the gross domestic product of the entire Polish border region eligible under the programme – a minuscule share compared to the much higher shares (between 1.5 and almost 4 per cent) granted in the form of Structural Funds to Cohesion Countries in the EU. It is also much less than the 4-per-cent threshold that will limit the flow of funds from the EU to the least developed member states during the programming period 2007–2013. However, it should be

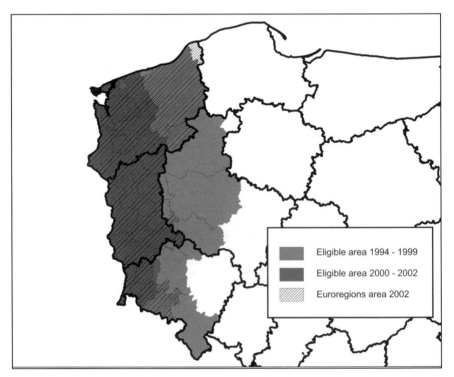

Map 14.2   Polish territories eligible for Poland-Germany Phare CBC grants

stressed that the inflow of EU funds triggered the release of more important domestic resources: overall spending was four times higher and amounted to € 1.108 billion during the years 1994–99.

At first the number of large investment projects carried out under PHARE-CBC grew with each subsequent year of the programme – up to 51 projects in 1997. Subsequently, it dropped to 23 (1999), that is to the level observed in 1995. Since the funds allocated to Poland's western border regions were similar for each year, the growth or decline in the number of projects entailed changes in the average value of the financial support awarded per project – from € 6.5 million in 1994 and € 0.9 million in 1997 to € 1.3 million in 1999, with an average value of € 1.5 million for the entire period. Municipal governments, who received 42 per cent of the funds implemented more than half of all investment projects, followed by other public institutions (26 per cent), particularly the regional branch offices of the General Directorate of National Roads and Motorways (formerly the General Directorate of Public Roads); Polish National Railways (PKP) was another beneficiary. The share of other organizations, such as the regional authorities (government and self-government), Euroregions, municipal unions, non-governmental organizations and universities, did not exceed 5 per cent for each category. A quite different picture emerges when the value of the project is taken into account: here public institutions were granted 31 per cent and voivodship offices 16 per cent of all allocated resources.

As in a traditional regional policy approach, over two thirds of outlays were directed to 'hard' projects in two sectors: transport and environmental protection. There were indeed good reasons for this approach as these two sectors greatly lacked funds for development, and backwardness constituted the biggest obstacle to Polish-German cooperation. The Polish-German border plays indeed a key role in trade with eastern Europe, and waste water treatment significantly affects the Nysa Luzycka and Odra (Oder) Rivers forming the natural boundary between the two countries.

In the mid–1990s, the poor quality of the border crossings was one of the most severe impediments of economic integration with Germany and the EU, and the CBC programme helped ease this difficulty considerably. Investments co-financed by PHARE-CBC allowed to increase the technical capacity of the crossing points at Poland's western border by 13,830 cars and 6,400 lorries per day – up to 20 per cent of the entire motor traffic passing through this border. The Polish-German border also channels half of the Polish pedestrian international traffic (totalling 278 million persons in 2000), almost half of all passenger traffic by road (45,4 million passenger cars and coaches), 57 per cent of the lorry traffic (out of 5,4 million lorries) and 55 per cent of the traffic in goods measured by value. PHARE resources were also used for the construction and modernization of roads leading to eleven border crossing points.

Four types of projects were implemented in the field of environmental protection: piped water systems; waste water treatment plants; garbage collection and recycling systems; and gazification projects. Forty-six joint plants for waste water treatment and sewage systems were constructed or improved, in some cases for the first time. On several occasions the very construction of such a system led to the creation of new housing and jobs. A joint undertaking by 13 municipalities which created a complex system of garbage collection, segregation and processing in Dlugoszyn is a 'widow-shop' case, demonstrating the value that can be added by intermunicipal cooperation to 'typical' projects co-financed by the EU.

Projects financing the improvement of human resources accounted for only 8 per cent of total spending. However, the greatest project in this category – the Polish part of the International University Viadrina in Frankfurt/Oder-Slubice – is a spectacular example of Polish-German and international cooperation in higher education. One third of all students come from Poland, and not only from the border region but from the entire country, while some German students live and study on the Polish side of the border, transforming the cities of Frankfurt and Slubice into true binational locations.

Projects undertaken under the label 'economic development' were in most cases related to tourist infrastructure, among them the construction of bicycle paths with the aim of creating a dense network on both sides of the border.

The social dimension of the CBC programme relies on almost 2,000 small projects undertaken in the period 1994–99 by municipalities and municipal institutions under the general umbrella of the Euroregions established along the Polish-German border (see Map 14.3). Sixty-one per cent of the 150 municipalities involved were part of a Euroregion. Two thirds of these projects – with budget of up to € 15,000 for a single project – were set up to promote cultural cooperation, such as mutual visits of school children, jointly organized cultural events, mutual visits of amateur theatre troupes, bands and orchestras.

Several surveys were conducted on both sides of the border to evaluate the impact of the PHARE-CBC programme in terms of cooperation between communities and individuals The overall picture can be called optimistic. Seventy-five per cent of the institutions that took part in the programme declared their participation to be important or very important for their overall activities. Almost all institutions replied that their expectations were met fully (58 per cent) or at least partially (40 per cent), and all those that were granted funds intended to apply for future projects. This share was much higher in localities which had benefited from PHARE-CBC (62.3 per cent, and only 44,7 per cent in those which had not). More interestingly, almost twice as many respondents in these localities indicated that the quality of their life had improved – a clear sign of the positive impact of large infrastructure projects.

Attitudes towards Neighbours

Research has shown that approximately three fourths of the surveyed Germans more or less regularly go shopping in Poland and nearly one third of the residents of Polish municipalities located near the border declare making shopping trips to Germany. Over 90 per cent of the Germans interviewed were of the opinion that goods in Poland are on the whole cheaper – certainly the main reason for such shopping trips. Germans mainly buy food (33 per cent), clothing and footwear (24 per cent),

Map 14.3   The German-Polish border region and Euroregions

cigarettes (16 per cent) and only rarely household appliances and cosmetics whereas Poles most frequently buy food (approximately 60 per cent), clothing and footwear (around 50 per cent), cosmetics (approximately 30 per cent), household appliances and home electronics, the latter three categories of products being unquestionably cheaper in Germany.

A considerably smaller number of the German respondents (approximately 20 per cent) buy services in Poland, and cultural contacts are slightly more frequent. One fourth of the interviewees take part in cultural events, although they do so irregularly. A similar share of the respondents take part in sports events. One fourth of the surveyed Germans declared spending their holidays in Poland, and one in ten respondents said he was maintaining social contacts with Poles. In Poland, approximately 45 per cent of the respondents living in municipalities near the border declared having friends in Germany. This favourable situation clearly results from the opening of numerous new border crossings. Only approximately 10 per cent of the interviewed Poles, and a similar share of Germans, complained of the inconvenience they had suffered from queuing at border inspection points. Distance from the border does not appear to restrict contacts.

More than half of the surveyed Germans knew about joint German-Polish programmes that had been implemented in their communities during the past ten years. At the same time, less than a half of the interviewees were aware that the municipality in which they live is part of a Euroregion, and only slightly more than one tenth of the respondents were able to quote the Euroregion's name. One third of the surveyed Germans were of the opinion that being part of a Euroregion brings benefits or significant benefits, while half of them mentioned both favourable and unfavourable effects. In Poland, approximately 30 per cent of the respondents pointed out advantages of being part of a Euroregion, while the majority did not have any firm opinion.

Residents of Polish municipalities identify most strongly with their country, the average answer being around 4.4 on a 5-point scale (where 5 is the strongest, and 1 the weakest identification), while Germans do so considerably less (an average of 3.3 on the same scale). Only 34 per cent of the German respondents, but approximately three fourths of the Polish interviewees, claimed a strong or very strong identification with their country.

In general, both the Polish and German inhabitants of the border regions can be divided into three broad categories. One third of the Germans are happy to have Polish neighbours, think that it is important to know the Polish language and have already started to learn it (Poles have been learning German for a long period, German being the most popular foreign language in Poland's western border region, contrary to the rest of the country where English leads by a large margin). Approximately one third of the Germans have a neutral attitude toward Poland, and the attitude of the remaining third is unfriendly. Similar proportions can be observed on the other side of the border – nearly 40 per cent of the respondents declared that they like or rather like Germans. Regardless of their declared feelings, both Germans (much more frequently) and Poles (less frequently) travel across the border for shopping trips. There is no doubt that the increasing frequency of such and other types of contacts can be attributed to the PHARE-CBC programme; its investments into border crossings has made it possible to relatively easily and quickly cross the Polish-German border.

Conclusions

The Polish-German border today resembles very much other internal EU borders between old member states, connecting rather than dividing two different and mutually complementary economic, cultural and social systems. It thus comes as no surprise that the support for Polish accession to the EU was strongest in Poland's western regions, which have benefited from EU-sponsored programmes for a period of ten years prior to full membership. Institutional support for enhancing cross-border cooperation and friendly neighbourhood relations has clearly been very important and come from a multitude of organizations: the European Union, national governments, regional authorities on both sides of the border, Euroregions and local governments. All these institutional partners were highly influential – and successful – in overcoming long-lasting prejudices and tensions between the citizens of the two nations. In addition, EU-financed PHARE programmes for cross-border cooperation appear to have been an excellent training ground preparing for full membership in the EU. Local governments and several organizations were de facto 'members' of the EU long before full accession. They had to follow the programming principles, fulfil certain requirements while preparing applications, and secure transparent accounting and honest reporting. After accession, western Poland was thus best prepared to take advantage of EU funds.

Finally, on the German side, Poland is no longer perceived as a distant country. The Polish consumer market has been a very profitable place for everyday shopping. It is also an example of private entrepreneurship and economic efficiency – features lacking in the former German Democratic Republic. Germans do not only gain through price differences but may also learn from their neighbours how to create economic success in a competitive world.

I argue that Polish-German cross-border cooperation can serve as a model for regions along the external borders of the enlarged European Union. There are indeed several similarities (and, of course, several differences, too) between the Polish-German and Ukrainian-Polish borders. One could dream that, in the near future, Ukraine's western regions will receive financial and institutional support similar to that offered to Poland's western regions ten years ago and that these efforts will be strongly supported by the European Union. Current changes in Ukraine suggest that this dream could definitely come true.

Chapter 15

Regional Development in Times of Economic Crisis and Population Loss: The Case of Germany's Eastern Border Region

Hans-Joachim Bürkner

A decade ago political discourses portrayed the German-Polish border region as an area ready to reap the benefits of European integration and of a neoliberal growth regime which would ensure that economic success and prosperity would radiate from a Berlin growth pole. In retrospect, and as early warnings voiced by critical observers predicted (Krätke 1996 and 1998), East Germany's peripheries have experienced a dramatic decline through deindustrialization and population loss. In the case of the German-Polish border region, there is now an increasing local demand for political action as EU enlargement has not substantially reduced the separating effects of the border. This contrasts considerably with political agendas at national and regional levels. While the German and Polish national governments plunged into a variety of hectic activities ahead of Poland's accession to the EU, an eerie calm has set in since May 2004, contributing to an atmosphere of disillusionment on both sides. Political initiatives for regional cooperation, ostensibly so intense in the run-up to enlargement are rather 'low-key'. Optimistic expectations for the near future as expressed by local politicians have given way to mild scepticism. In particular, politicians' dreams of a new sense of transboundary 'regionalism' seem to have been unrealistic options given the present situation. Many local actors have grown apathetic at the same time that fears of having 'lost out in the enlargement game' are quite present, especially on the German side (see Dürrschmidt 2002). Local decision-makers have finally become aware of the underlying and drastic changes that have taken place not only in international relationships but also with regard to regional development along the border. Thus, while being exposed to intensified interregional competition, the border region will see a gradual reduction of its eligibility for EU structural funds, so far a major source for financing regional development in Brandenburg (Karl 2002, p. 209).[1] Although economic prospects for the coming years indicate that Germany

---

1 Brandenburg will lose its Objective 1 status in 2006 and, as a result, many of the subsidies and resources vital to the stabilization of the border region.

and the accession states will enjoy economic benefits from EU enlargement, it has been pointed out that already prosperous regions will profit far more than peripheries (Krätke and Borst 2004). Given long-term trends of structural decline, especially low investment, high rates of unemployment and outmigration, the border region risks being caught in a permanent development trap.

Structural Disparities between the German and Polish Parts of the Border Region

Recent processes of marginalization have concealed the fact that from the very beginning, structural 'points of departure' for the German-Polish border region have been unfavourable for both sides. In addition to regional effects of globalization, negative effects of societal transformation such as unemployment, outmigration and a low participation rate in regional labour markets have left their imprints on the border region, yet in different ways, leading to structural disparities between the German and Polish subregions, which at this point appear irreversible. East Brandenburg achieved a higher level of wealth solely because of massive financial transfers from the federal government and the EU as well as rapid integration into the German welfare state. In contrast, western Poland embarked on a rather grinding and risky path of development which resulted in more economic dynamism (Krätke 1998, p. 252).

Map 15.1   The German-Polish Euroregion Pro-Europa Viadrina

Development Trends in East Brandenburg

West of the border, the transition towards a market economy has resulted in a protracted process of deindustrialization. Mass unemployment and migration to Berlin and western Germany have thinned out regional human capital. Former employees of large production plants of the GDR semiconductor industry at Frankfurt/Oder, for example, turned their backs on the region in the early 1990s, never to return (Matthiesen and Bürkner 2004, p. 83). Globalization has made this process irreversible, since the rapid devaluation of technological skills within new knowledge-based industries and services has reduced chances of labour market reintegration for the remaining qualified workforce. Recent attempts to revitalize semiconductor production through forming a small cluster of highly specialized small enterprises and a business incubator at Frankfurt/Oder have hardly affected unemployment levels, as the professional skills required cannot be found locally any more. At the same time large transnational corporations (TNCs) have shown almost no interest in the region.[2] From the mid–1990s on, rising wages and decreasing subsidies have encouraged enterprises to relocate both labour-intensive and knowledge- and capital-intensive production directly to eastern Europe, thus bypassing East Germany. Minor exceptions apart (Krätke 1998 and 1999), there have been no research and development activities in the German-Polish border region. The classical definition of economic peripheralization thus fully applies.

Indeed, despite some promising beginnings at Frankfurt/Oder, the 'new economy' has not offered a viable development strategy for the border region, at least not one that might stave off structural decline. In addition to a qualified labour force and specialized research facilities, firms operating within the knowledge economy (IT services, semiconductor production etc.) require particular locational qualities, such as the presence of cultural amenities appealing to members of the 'creative classes' (Florida 2002, p. 218), of which East Brandenburg has little to offer. Except for the Technical University of Cottbus, there is very little research and development infrastructure that meets the needs of 'global players'. The only major initiative along these lines were plans, now obsolete because the necessary venture capital never materialized, to establish a large microchip factory at Frankfurt/Oder, hardly anything more than an extended high-tech workbench (Kühn 2004. p 263).

Another aspect of the 'New Economy' development doctrine operating in the German-Polish border region was the notion of promoting networks of locally based small and medium-sized enterprises (SMEs). These strategies developed only slowly in spite of governmental funding, extensive consulting advice and the establishment of business incubators. One such incubator, a rather singular case, is a cluster of microchip producers assembled around a state-funded research institute of semiconductor physics at Frankfurt/Oder. This institute was able to sustain its operations for more than a decade but received little political support at the local

---

2 A modest number of extended workbenches of the *maquiladora* type in traditional sectors depend on cheap labour and government subsidies, rather variable factors over time.

level as it was only poorly integrated into Frankfurt's urban region and exclusively oriented towards global markets. Effects for the region as well as for cross-border cooperation could only be expected to be marginal (Matthiesen and Bürkner 2004, p. 83).

Neither has transboundary cooperation in more traditional economic sectors led to satisfactory results. While big companies have better access to Polish markets since the EU enlargement, the region's SMEs, formerly handicapped by different trade structures and business procedures (Bürkner 2002), still face substantial barriers: communication problems and slow developing trust continue to form the single most important obstacle to cooperation and are often exacerbated by local hostility towards foreigners (Krätke 1998, p. 258; 2004, p. 87). Moreover, except for the Szczecin agglomeration, attractive business partners in Poland are usually operating in more distant industrial core regions further east, such as Poznan, Wrocław and Warsaw. Firms located on the German side of the border region only play a marginal role in German-Polish trade (Krätke 2004, p. 84), which is dominated by partnerships between enterprises with head offices in western or southern Germany and central or, to a lesser extent, western Poland.

As East Brandenburg has experienced a steady demographic decline since the late 1990s, loosing its most skilled labour force, and large numbers of young people continue to leave the region for more prosperous areas elsewhere in Germany, local governments are facing new problems of planning and structural development to cope with huge housing vacancies and obsolete urban infrastructure as a result of 'shrinking cities', a widespread phenomenon in East Germany. A recent government programme for the physical transformation of towns and cities ('Stadtumbau Ost'), for instance, has provided support for the demolition or downsizing of vacant buildings, primarily on large housing estates built during the socialist period. Though additional funds were allocated for improving social infrastructure and other non-physical assets, and local governments seeking grants were required to develop Integrated Urban Development Strategies (Kabisch et al. 2004, p. 142), major cultural, social and economic issues of urban decline have been ignored (Institut für Stadtforschung und Strukturpolitik 2004, p. 16). The main beneficiaries of the programme thus appear to have been local housing associations who were facing bankruptcy because of vacancies and operational losses (Bernt 2005, p. 126).[3] Many residents have been forced to relocate to new neighbourhoods, a move whose long-term social and economic effects remain to be studied. All this strongly suggests insufficient recognition of basic local and regional development problems by the regional government.

---

3   Recent proposals to establish cross-border housing associations to overcome housing shortages in Polish border cities or to allocate vacant flats in Frankfurt/Oder to residents of its twin city Slubice, as suggested by the mayor of Frankfurt/Oder, have found little resonance with, if not encountered outright hostility from the German local population (Klesmann and Kursa 2003).

Social and cultural issues as reflected in everyday culture or in sentiments and the general outlook of people from different social milieux are indeed absent from political debates. Whereas official policy has been emphasizing the promotion of transboundary communication within the framework of INTERREG and PHARE-CBC programmes, fear of increased labour market competition and Polish immigration merges with racist undercurrents on the German side where some social milieux tend to seal themselves off from their Polish neighbours who, after initial curiosity, now manifest increasing scepticism (Bürkner 2002). Critics contend that years of symbolic politics have not taken sufficient care of people's actual needs and mentality. Many German citizens in the border region have adopted a rhetoric of general decline and, in the name of local identity, now often reject attempts from the outside to change the present situation as interference into the 'internal affairs' of their community.[4]

Development Trends in western Poland

Although Poland's postsocialist transformation is often characterized as economic 'shock therapy', deindustrialization has hit western Poland much less than East Germany. Local industries there have survived through integration into the regional and national value chain. Most of them are operating in traditional branches, such as food, clothing and timber industries, and often enjoy a well-established market presence, though without much effect on employment levels (Krätke and Borst 2004, p. 94; Stein, 1997). Companies in technology-based and innovative sectors, on the contrary, are almost absent (Stryjakiewicz 2004, p. 125) and are generally located in more dynamic urban centres in central Poland, such as Poznan. A growing number of small businesses as well as services of a largely informal character (Stryjakiewicz, 2004, p. 122) have at least partially compensated job losses in other economic sectors, and flourishing bazaar markets in the 1990s have contributed to an overall neutral migratory balance, by attracting job-seekers from central and eastern Poland. The region has thus maintained a pool of low-skilled labour.

Foreign direct investment to the region has been mainly directed towards 'low-tech' sectors. Large furniture makers from western Europe, for instance, have established production facilities of an export-processing character, taking advantage of low wages for semi-skilled labour (Krätke and Borst 2004, p. 16). A classical case of 'leapfrogging', these investments have no positive effect on the German part of the border region. The only microregional type of factor exchange is due to a small number of binational enterprises, also in the furniture industry. These have set up production sites on the Polish part of the border region and are selling their products on the German side, mainly through factory outlets – one of the few profitable forms

---

4   Research on border milieux in the twin city of Guben-Gubin conducted by the Leibniz Institute for Regional Development and Structural Planning, Erkner, has often recorded statements such as 'Our town is going to die', which are now quite common in East Brandenburg (Dürrschmidt 2002).

of cross-border business below the level of transnational corporations (Matthiesen and Bürkner 2002).

Future Prospects

In short, whereas East Brandenburg continues to suffer from stagnation and even negative growth rates, western Poland has maintained a relatively strong position within the national context, as highlighted in particular by the agglomeration of Szczecin and its remarkable concentration of economic potential (Krätke and Borst 2004, p. 86). As a consequence, gaps in wealth between areas west and east of the border have been reduced since the mid–1990s (Krätke 1998, p.253). Future prospects, however, look bleak not only for the German but also for the Polish part of the border region. Preliminary evidence of the effects of the 2004 EU enlargement indicates that it is already prosperous regional economies of the 'old' member states, such as economic cores with strong economic, technological and institutional resources in the western part of Germany, and regional economic cores in Central European states which will benefit the most from more intense economic cooperation (Krätke 2004), and that large economic centres will exert an even stronger pull on capital and labour. In Poland, positive effects are expected for agglomerations such as Warsaw as well as for strong regions in the western part, such as that of Poznan (Krätke 2004; Gorzelak 1998; Barjak 2001).

Chances for long-term economic growth in the German-Polish border region, on the other hand, remain uncertain. As more optimistic development scenarios have disappeared, a much reduced vision claims future benefits for the region from functions linked to transit between the metropolitan areas of Berlin and Warsaw or from spill-over effects within a broadened European 'region of competence'. However, existing railway and road links do not allow for the speed necessary to meet business needs – many journeys by rail between Berlin and Polish cities today take longer than they did in the 1920s or 1930s (Dyckhoff 2004). Moreover, the German and Polish governments have different strategies when it comes to the development of major transport routes, with the Germans favouring East-West transboundary connections, while the Poles intend to improve North-South routes within their country (Bürkner and Uhrlau 2002). Other spatial planning concepts appear to be more consensual. Since 2000 the German-Polish Intergovernmental Commission has moved from a microregional and mainly corridor-oriented concept to a more growth-pole oriented paradigm similar to the polycentric development model favoured by the European Union. One instance of this is the project known as the 'German-Polish House' (Selke 2004), which transcends the narrow strip of the economically weak border region in favour of an extended regional space including major regional centres in East Germany and western Poland. On the local level, binational cooperation in the management of EU initiatives, as Grzegorz Gorzela (this book) correctly points out, has promoted interaction of public actors and organizations emanating from civil society and helped building institutional capacities at this level. In practical terms, the failure to develop adequate policy responses, particularly on the German side of the border, should encourage stakeholders within the region to reassess local resources and potentials for action.

Dealing with the Governance Question: Building Up Local Capacities

How can the EU integration process be influenced so that positive effects for the German-Polish border region can be achieved? What policy and governance options would allow to improve cross-border communication? Presently, we can identify at least four fields of action where steering capacities in terms of cross-border cooperation could be developed: 1) the management of EU structural policy; 2) the development of more responsive economic policies at state level; 3) more effective cooperation in spatial planning; and 4) the mobilization of local capacities for self-organization and the promotion of cross-border learning processes. All these governance options are, of course, closely interrelated and deserve some elaboration.

There is no doubt that even after direct structural fund subsidies for East Germany decrease in volume (see DIW 2001), EU incentives for cross-border cooperation through the INTERREG initiative will be direly needed, especially to improve local infrastructure. Local governments in western Poland, for instance, lack financial resources to pay for better road links and urban services. Of equal importance is the promotion of regional economic activities on both sides of the border. Instead of waiting for investors, local policy should concentrate on promoting local small and medium-sized enterprises through the judicious and strategic use of EU funds. Past experience has shown that local governments prefer to apply for INTERREG funds in view of financing cultural and other projects with minor structural effects, but which have helped improve 'person-to-person' contacts within the region. This focus need not exclude projects that promote transboundary enterprise networks or develop consultancy services for local enterprises. Indeed, INTERREG funds for local economic development have been, and will continue to be, easily available and some have still not yet been spent. Finally, with Poland's accession to the EU, the management of cross-border resources provided by the EU has been considerably simplified – a further incentive for more sustainable cooperation projects.

However, better use of EU funds for cross-border cooperation must receive more support from the state government. This will require a major rethinking of economic policy, for example, on the part of the regional government of Brandenburg. As described above, regional development strategies for East Brandenburg have been largely unsuccessful, focusing primarily on the promotion of industrial locations for traditional branches as well as for a small number of high-tech SME clusters, often accompanied by business incubators with no local links (Krätke and Scheuplein 2001). On the other hand, clusters of enterprises in high-tech production branches that emerged spontaneously have been virtually ignored. An alternative strategy could therefore be to encourage incipient entrepreneurial development in the field of knowledge-based production and services. Furthermore, creating facilities for education and vocational training suitable for the specific needs of the knowledge economy might, at least in some areas, help develop human capital in the long run. However, expectations that such policies alone might give rise to a favourable investment climate and new economic futures may be unrealistic for eastern Germany where even good local infrastructure combined with research and educational facilities have seldom resulted in improved economic outcomes (Franz 2004, p. 119).

Transboundary cooperation in spatial planning, the third governance option, has so far not met expectations. Since 1990 several German-Polish spatial planning institutions have been established but with varying results (see Scott 1998). At the national level, the German-Polish Intergovernmental Commission for the Development of the Border Region attempted to establish a joint agenda in 2001 and 2002 (Bürkner and Uhrlau 2002). However, its objectives were too diffuse and partially incompatible, reflecting different strategic outlooks. By using guidelines defined in the European Spatial Development Concept (Selke 2004), the subsequent concept of the 'German-Polish House' seems to have a better chance of harmonizing national planning through linking the border region more directly and more effectively to European core regions (Weislo 2004). But, once more, prevailing centralist notions of planning and objections to the idea of harmonization-cum-'integration' voiced by national actors, especially in Poland, might prove too high a hurdle. It therefore remains unclear to what extent regional and local planning agencies will participate in this exercise. Independent cooperation initiatives underwritten by East Brandenburg regional planning institutions, for example, have not been very successful due to lack of political support at the national level. More political backing for joint planning efforts will be required as integration within the EU-25 proceeds.

This leads to the fourth governance issue, that of improved local capacities for action through the initiation and support of learning processes linking actors on both sides of the border. Local governance, defined as an interplay of formal state organizations, private actors and the 'third' sector, can already build upon existing intermediary organizations which have developed a variety of cross-border relationships and communicative mechanisms hardly noticed by the public. Local associations, clubs, private initiatives and non-governmental organizations are firmly rooted in everyday life and enjoy a certain legitimacy among the local population (Matthiesen and Dürrschmidt 2002). They are not based on an abstract, exogenous political idea, but on interests shared by individual citizens who pursue, for example, hobbies or sports activities or are nurturing the regional cultural heritage. Successful cross-border communication relies on the gradual and 'organic' building of trust rather than on symbolic political efforts to create a sense of common regional identity. Time and patience are required in order to overcome mental barriers to enlargement, such as the tenacious 'Europhobia' embedded in local culture west of the border – an issue that has often been forgotten in debates on the EU and its future development.

Intermediary organizations can also assume the role of mediators, reconcile everyday culture with formal levels of politics and the economy, and exercise a stabilizing influence. In order to do this effectively, they require more active support from local, regional and national Western governments. Such a 'learning border region' (see Matthiesen and Bürkner 2001) might be more successful in responding to the present challenges of mental peripheralization and economic depression.

Bibliography

Aghrout, Ahmed (2000) *From Preferential Status to Partnership: The Euro-Maghreb Relationship* (Aldershot: Ashgate).

Agnew John and Stuart Corbridge (1989) 'The New Geopolitics: The Dynamics of Geopolitical Disorder', in Ron J. Johnston and Peter J. Taylor (eds) *A World in Crisis?: Geographical Perspectives* (Oxford: Basil Blackwell).

Anderson, Malcolm and Eberhard Bort (2001) *The Frontiers of the European Union* (Houndmills: Palgrave).

Anderson, Sean (2004) 'Bystroye Canal and the Danube River Delta: An International Controversy', *Global Built Environment Review*, 4, 2: 7–10. Available at <www.edgehill.ac.uk/gber/pdf/vol4/issue2/Commentary2.pdf> [last access on 1 October 2005].

Ankler, Géza (1997) 'Nemzetközi tárgyilagosság és méltányosság', *Valóság*, XL (8): 1–32.

Antonsich, Marco (2002) 'Regionalization as a Way for Northern "Small" Nations to Be Heard in the New EU', in Finnish Institute of International Affairs (FIIA/UPI), *The New North of Europe*, Policy Memos of the Finnish Institute of International Affairs (UPI), Helsinki, 1–4.

Apap, Joanna and Angelina Tchorbadjiyska (2004) *What about the Neighbours? The Impact of Schengen along the EU's External Borders*, Working Document No. 210 (Brussels: Centre for European Policy Studies). Available at <shop.ceps.be/BookDetail.php?item_id=1171> [last access on 1 October 2005].

Balcsók, István and László Dancs (2001) 'A határon átnyúló kapcsolatok lehetőségei Magyarország egyik leendő schengeni határán', in Gábor Dormány, Ferenc Kovács, Márton Péti and János Rakonczai (eds) *A földrajz eredményei az új évezred küszöbén. A Magyar Földrajzi Konferencia tudományos közleményei* (Szeged: SZTE TTK Természeti Földrajzi Tanszék).

Barnier, Michel (2001) 'Cohesion in an Enlarged EU', speech delivered at the Second European Cohesion Forum, 21 May 2001, SPEECH/01/230. Available at <europa.eu.int/rapid/pressReleasesAction.do?reference=SPEECH/01/230&format=HTML&aged=1&language=EN&guiLanguage=fr> [last access on 24 November 2005].

Baranyi, Béla, (1999) 'A "periféria perifériáján" – a határmentiség kérdőjelei egy vizsgálat tükrében az Északkelet-Alföldön', *Tér és Társadalom*, 4: 17–44.

———— (ed.) (2000) *A határmentiség kérdőjelei az Északkelet-Alföldön* (Pécs: Centre for Regional Studies of the Hungarian Academy of Science).

——— (2002a) 'Euroregionális szervezetek és új interregionális szerveződések Magyarország keleti államhatárai mentén', *Magyar Tudomány*, 11 (2002): 1505–18.

——— (2002b) *Before Schengen – Ready for Schengen. Euroregional Organisations and New Interregional Formations at the Eastern Borders of Hungary*, Discussion Papers 38 (Pécs: Centre for Regional Studies of the Hungarian Academy of Science).

——— (2003) 'Euroregional organisations and formations on the eastern borders of Hungary', *European Spatial Research and Policy*, 1: 85–94.

——— (2004) *A határmentiség dimenziói. Magyarország és keleti államhatárai* (Budapest and Pécs: Dialóg Campus Kiadó).

——— (ed) (2005) *Hungarian-Romanian and Hungarian-Ukrainian Border Regions as Areas of Co-operation along the External Borders of Europe* (Pécs: Centre for Regional Studies).

Baranyi, Béla, István Balcsók, László Dancs and Barna Mező, *Borderland Situation and Peripherality in the Northeastern Part of the Great Hungarian Plain*, Discussion Papers, 31 (Pécs: CRS of HAS, 1999).

Barjak, Franz (2001) 'Regional Disparities in Transition Economies: A Typology for East Germany and Poland', *Post-Communist Economies*, 13 (3): 289–311.

Baronin Jr, Anatoliy S. (2001) 'Border Closed?', *Central Europe Review*, 3 (11). Available at <www.ce-review.org/01/11/baronin11.html> [last access on 5 October 2005].

Bauman, Zygmunt (1990) Modernity and Ambivalence, in Mike Featherstone (ed), *Global Culture: Nationalism, Globalization and Modernity* (London: SAGE Publications).

——— (1997) *Postmodernity and its Discontents* (Cambridge: Polity Press).

Beluszky, Pál (1995) 'Közép-Európa œ Merre Vagy?', *Földrajzi Közlemények*, CXIX (XLIII) (3-4): 223–32.

Benediktov, Kirill (2002) *Russian Security in the Context of EU Enlargement* (Moscow: Moscow School of Political Studies Internet Project).

Benhabib, Seyla (1996) 'Toward a Deliberative Model of Democratic Legitimacy', in Seyla Benhabib, *Democracy and Difference: Contesting the Boundaries of the Political* (Princeton: Princeton University Press).

Berg, Eiki (ed.) (2001) *Negotiating Borders of Multiple Meanings* (Tartu: Peipsi Centre for Transboundary Cooperation).

Bernt, Matthias (2005) 'Stadtumbau im Gefangenendilemma', in Christine Weiske, Sigrun Kabisch and Christine Hannemann (eds), *Kommunikative Steuerung des Stadtumbaus. Interessengegensätze, Koalitionen und Entscheidungsstrukturen in schrumpfenden Städten* (Wiesbaden: VS Verlag für Sozialwissenschaften).

Biroul National de Statistica (2004) *Population Figures for Moldova* (Chisinau: Biroul National de Statistica).

Blandy, Sarah, Diane Lister, Rowland Atkinson and John Flint (2003) *Gated Communities: A Systematic Review of the Research Evidence* (Summary) (London, Office of the Deputy Prime Minister).

Blatter, Joachim and Norris Clement (2000) 'Cross-Border Cooperation in Europe: Historical Development, Institutionalization and Contrasts to North America', *Journal of Borderlands Studies*, 15 (1):15–53.

Bognár, Károly (1993) 'A biztonság és biztonságpolitika hazánkban', *Társadalmi Szemle*, XLVIII (1): 39–48.

Bradley, Harriet (1997) *Fractured Identities: Changing Patterns of Inequality* (Cambridge: Polity Press).

Brown, David (2002) 'Storming the Fortress: The External Border Regime in an Enlarged Europe', in Hilary Ingham and Mike Ingham (eds), *EU Expansion to the East* (Cheltenham: Edward Elgar).

Browning Christopher (2001) *The Construction of Europe in the Northern Dimension*, Working papers of the Copenhagen Peace Research Institute, 39/2001 (Copenhagen: Copenhagen Peace Research Institute) (Contact through <www.copri.dk>).

Bruchis, Michael (1994) *The Republic of Moldavia: from the Collapse of the Soviet Empire to the Restoration of the Russian Empire* (Boulder: East European Monographs).

Bucken-Knapp, Gregg and Michael Schack (2001) 'Borders Matter, But How?', in Gregg Bucken-Knapp and Michael Schack (eds), *Borders Matter: Transboundary Regions in Contemporary Europe* (Aabenraa: Danish Institute of Border Region Studies).

Bukkvoll, Tor (2004), 'Private Interests, Public Policy: Ukraine and the Common Economic Space Agreement', *Problems of Post-Communism*, 51 (5): 11–22.

Burdack, Joachim, Frank-Dieter Grimm and Leo Paul (eds) (1998) *The Political Geography of Current East-West Relations*, Beiträge zur Regionalen Geographie, 47 (Leipzig: Institut für Länderkunde).

Bürkner, Hans-Joachim (2002) 'Border Milieux, Transboundary Communication and Local Conflict Dynamics in German-Polish Border Towns: The Case of Guben and Gubin', *Die Erde*, 133: 339–51.

Bürkner, Hans-Joachim and Andreas Uhrlau (2002) 'Spatial Planning and Regionalisation in the German-Polish Border Area: Harmonised Strategies or Emerging Lines of Conflict?' *European Spatial Research and Policy*, 9: 157–73.

Carter, Francis W. (1995) 'Közép-Európa: valóság vagy Földrajzi fikció?', *Földrajzi Közlemények*, CXIX (XLIII) (3-4): 232–49.

Catellani, Nicola (2002) 'Placing the Northern Dimension in Post-Enlargement EU-Russian Relations', in Finnish Institute of International Affairs (FIIA/UPI), *The New North of Europe*, Policy Memos of the Finnish Institute of International Affairs (UPI), Helsinki, 15–18.

Centre for International Relations (2002) *The EU's Eastern Dimension – An Opportunity for Idée Fixe of Poland's Policy?* (Warsaw: Centre for International Relations).

Chomette, Guy-Pierre (2002) 'La Moldavie repoussée vers l'Est', *Le Monde diplomatique*. Janvier.

Cilinca, Victor (2004) 'Cooperare Transfrontaliera', *Viata Libera*, 2 November.

Cimosewicz, Wlodzimierz (2003) 'The Eastern Dimension of the European Union: The Polish View', speech given at the 'EU Enlargement and Neighbourhood Policy Conference', 20 February 2003, Warsaw. Available at <www.polishembassy.ca/news_details.asp?nid=85> [last access on 22 November 2005].

Comaroff, Jean and John Comaroff John (2002) 'Millenium Capitalism: First Thoughts on a Second Coming', *Public Culture*, 12 (2): 291–344.

Commission of the European Communities (1997a) *For a Stronger and Wider Union. Agenda 2000, vol. I. Communication of the Commission*, COM(97) 2000 final. Available at <www.ena.lu/europe/european-union/commission-agenda-2000-stronger-union-1997.htm> [last access on 7 October 2005].

———— (1997b) *Establishing a partnership between the European Communities and their Member States, of the one part, and the Russian Federation, of the other part, Official Journal of the European Communities*, L 327. Available at <europa.eu.int/comm/external_relations/russia/pca_legal/> [last access on 6 October 2005].

———— (1998). *Communication from the Commission. A Northern Dimension for the Policies of the Union*, COM (1998) 0589 final. Available at <europa.eu.int/comm/external_relations/north_dim/doc/com1998_0589en.pdf> [last access on 6 October 2005].

———— (1999) *European Spatial Development Perspective. Towards Balanced and Sustainable Development of the Territory of the European Union*, (Luxembourg: Office for Official Publications of the European Communities). Available at <europa.eu.int/comm/regional_policy/sources/docoffic/official/reports/pdf/sum_en.pdf> [last access on 29 September 2005].

———— (2000) *Communication from the Commission to the Council and the European Parliament. On a Community Immigration Policy*, COM(2000) 757 final. Available at <europa.eu.int/eur-lex/lex/LexUriServ/LexUriServ.do?uri=CELEX:52000DC0757:EN:HTML> [last access on 12 October 2005].

———— (2001a) *Euro-Med Partnership. Regional Strategy Paper 2002-2006 and Regional Indicative Programme 2002-2004*. Available at <www.europa.eu.int/comm/external_relations/euromed/rsp/rsp02_06.pdf> [last access on 29 September 2005].

———— (2001b) *European Governance: A White Paper*, COM(2001) 428 final. Available at <europa.eu.int/eur-lex/en/com/cnc/2001/com2001_0428en01.pdf> [last access on 29 September 2005].

———— (2001c) *Country Strategy Paper 2002-2006. National Indicative Programme 2002-2003. Ukraine*. Available at <europa.eu.int/comm/external_relations/ukraine/csp/02_06en.pdf> [last access on 5 October 2005].

———— (2001d) *Proposal for a Council Directive on the Conditions of Entry and Residence of Third-Country Nationals for the Purpose of Paid Employment and Self-Employed Economic Activities*, COM(2001) 386 final. Available at <europa.eu.int/eur-lex/lex/LexUriServ/site/en/com/2001/com2001_0386en01.pdf> [last access on 12 October 2005].

—— (2002) *Communication from the Commission to the Council and the European Parliament*, COM(2002) 703 final. Available at <europa.eu.int/eur-lex/lex/LexUriServ/site/en/com/2002/com2002_0703en01.pdf> [last access on 12 October 2005].

—— (2003a) *Communication from the Commission to the Council and the European Parliament: Wider Europe – Neighbourhood: A New Framework for Relations with our Eastern and Southern Neighbours*, COM(2003) 104 final. Available at <europa.eu.int/eur-lex/lex/LexUriServ/site/en/com/2003/com2003_0104en01.pdf> [last access on 29 September 2005].

—— (2003b) *Communication from the Commission: Paving the Way for a New Neighbourhood Instrument*, COM (2003) 393 final. Available at <europa.eu.int/eur-lex/lex/LexUriServ/site/en/com/2003/com2003_0393en01.pdf> [last access on 29 September 2005].

—— (2003c) *Ukraine. National Indicative Programme, 2004-2006*. Available at <europa.eu.int/comm/external_relations/ukraine/csp/ip03_04_08.pdf> [last access on 5 October 2005].

—— (2003d) 'Action Plan on Justice and Home Affairs concerning Ukraine', *Official Journal of the European Union*, C77: 1–5. Available at <europa.eu.int/eur-lex/pri/en/oj/dat/2003/c_077/c_07720030329en00010005.pdf> [last access on 5 October 2005].

—— (2003e) *Commission Working Document. The Second Northern Dimension Action Plan, 2004-2006*, COM(2003) 343 final. Available at <europa.eu.int/eur-lex/lex/LexUriServ/site/en/com/2003/com2003_0343en01.pdf> [last access on 6 October 2005].

—— (2004a) *Communication from the Commission, European Neighbourhood Policy. Strategy Paper*, COM (2004) 373 final. Available at <europa.eu.int/eur-lex/lex/LexUriServ/site/en/com/2004/com2004_0373en01.pdf> [last access on 29 September 2005].

—— (2004b) *Proposal for a Regulation of the European Parliament and the Council Laying Down General Provisions Establishing a European Neighbourhood and Partnership Instrument*, COM (2004) 628 final. Available at <europa.eu.int/eur-lex/lex/LexUriServ/site/en/com/2004/com2004_0628en01.pdf> [last access on 29 September 2005].

—— (2004c) *Proposal for a Council and Commission Decision on the conclusion of the Protocol to the Partnership and Cooperation Agreement between the European Communities and their Member States, of the one part, and the Russian Federation, of the other part, to take account of the accession of the Czech Republic, the Republic of Estonia, the Republic of Cyprus, the Republic of Latvia, the Republic of Lithuania, the Republic of Hungary, the Republic of Malta, the Republic of Poland, the Republic of Slovenia and the Slovak Republic to the European Union*, COM (2004) 292 final - CNS 2004/0087. Available at <europa.eu.int/eur-lex/lex/LexUriServ/site/en/com/2004/com2004_0292en01.pdf> [last access on 7 October 2005].

———— (2004d) *Communication from the Commission to the Council and the European Parliament on relations with Russia*, COM(2004) 106 final. Available at <europa.eu.int/eur-lex/lex/LexUriServ/site/en/com/2004/com2004_0106en01.pdf> [last access on 6 October 2005].

———— (2004e) *Communication from the Commission. European Neighbourhood Policy. Strategy Paper*, COM (2004) 373 final. Available at <europa.eu.int/eur-lex/lex/LexUriServ/site/en/com/2004/com2004_0373en01.pdf> [last access on 6 October 2005].

———— (2004f) *Communication from the Commission to the Council on the Commissions Proposal for Action Plans under the European Neighbourhood Policy (ENP)*, COM(2004) 795 final. Available at <europa.eu.int/eur-lex/lex/LexUriServ/site/en/com/2004/com2004_0795en01.pdf> [last access on 6 October 2005].

———— (2004g) *Communication from the Commission to the Council and the European Parliament. Building Our Common Future Policy Challenges and Budgetary Means of the Enlarged Union 2007-2013*, COM(2004) 101 final 2. Available at <europa.eu.int/eur-lex/lex/LexUriServ/site/en/com/2004/com2004_0101en02.pdf> [last access on 6 October 2005].

Commission of the European Communities–Economic and Social Committee (2002) 'Opinion of the Economic and Social Committee on the "Euro-Mediterranean Partnership – Review and Prospects Five Years On"', *Official Journal of the European Communities*, 2002/C32/24: 117–26.

Commission of the European Communities–External Relations (2003) *The Euro-Mediterranean Free-Trade Area. Set to Become the World's Biggest Marketplace.* Available at <www.europa.eu.int/comm/external_relations/euromed/free_trade_area.htm> [last access on 29 September 2005].

Constitutional Watch (2004) 'Moldova', *East European Constitutional Review*, 1-2: 29–33.

Council of the European Union (1996) 'Council Regulation (EC) No 1488/96 of 23 July 1996 on financial and technical measures to accompany (MEDA) the reform of economic and social structures in the framework of the Euro-Mediterranean partnership', *Official Journal of the European Communities, 30 July, L 189: 1–9. Available at <europa.eu.int/eur-lex/lex/LexUriServ/LexUriServ.do?uri=CELEX:31996R1488:EN:HTML> [last access on 7 October 2005].*

———— (1999a) 'Common Strategy of 11 December 1999 on Ukraine', *Official Journal*, L331: 1–9. Available at <www.eclc.gov.ua/new/html/eng/7/clc_2_strategy.html> [last access on 5 October 2005].

———— (1999b). Common strategy of the European Union on Russia, 10810/99. Available at <register.consilium.eu.int/pdf/en/99/st10/10810-c1en9.pdf> [last access on 6 October 2005].

———— (1999c) Council Regulation (EC, Euratom) No 99/2000 of 29 December 1999 concerning the provision of assistance to partner States in Eastern Europe and Central Asia, Official Journal of the European Union, L 012, 18 January 2000:

1–9. Available at <europa.eu.int/eur-lex/lex/JOHtml.do?uri=OJ:L:2000:012: SOM:EN:HTML> [last access on 9 October 2005].

———— (2000) *Action Plan for the Northern Dimension with external and cross-border policies of the European Union 2000-2003*, 9401/00. Available at <register. consilium.eu.int/pdf/en/00/st09/09401en0.pdf> [last access on 6 October 2005].

———— (2001), Full Report on Northern Dimension Policies, Document 9804/01 NIS 43/COEST 16/PESC 225, 12 June 2001. Available at <register.consilium. eu.int/pdf/en/01/st09/09804en1.pdf> [last access on 29 September 2005].

Cranganu, Nicoleta (2003) 'Filiala din Cahul a Universitatii "Dunarea de Jos" are probleme', V*iata Libera*.

Cronberg, Tarja (2000) 'Euroregions in the Making: The Case of Euroregio Karelia', in Pirkkolissa Ahponen and Pirjo Jukrainen (eds), *Tearing Down the Curtain, Opening the Gates: Northern Boundaries in Change* (Jyväskylä: SoPhi Academic Press).

———— (2001) *Europe Making in Action: Euregion Karelia and the Construction of EU-Russian Partnership*, paper presented at the 'Think-Tank Seminar' on the 'Northern Dimension and Future of Barents Euro-Arctic Co-operation', 14-17 June, Björkliden, Sweden.

Cronberg, Tarja and Valery Shlyamin (1999) 'Euregio Karelia – A Model for Cooperation at the EU External Borders', *Crossing the Borders in the Northern Dimension.* (Oulu: publisher).

Crowther, William (1998) 'Ethnic Politics and the Post-Communist Transition in Moldova', *Nationality Papers*, 26 (1): 150-51.

———— (2000) *Between Transdniestria and Snake Island: The Evolution of Romanian Foreign Policy Toward Ukraine*, paper delivered at the "Ukraine and Central Europe: Multi-level Networks and International Relations" conference of the Harvard Ukrainian Studies Center and the Woodrow Wilson Center, at the Woodrow Wilson Center, Washington DC, 18-19 May.

Cucu, Vasile and Gheorghe Vlasceanu (1991) *Insula Serpilor* (Bucuresti: Viata Romaneasca).

Deák, Péter (1997) 'Uniós biztonságpolitika: egységes, közös vagy egyeztetett?', *Európai Tükör*, II (1): 47–59.

———— (2004) 'EU- és NATO-kapcsolatok - a megegyezések és viták tükrében', *Európai Tükör*, 9 (1): 24–45.

Decker, Ulrike (2002) 'The EU's New Eastern Dimension: A Gateway to a New North-Eastern Cooperation', in Finnish Institute of International Affairs (FIIA/ UPI) (ed.), *The New North of Europe*, Policy Memos of the Finnish Institute of International Affairs (UPI), Helsinki, 23–6.

Deichmann, Thomas, Sabine Reul and Slavoj Zizek (2002) 'About War and the Missing Center in Politics', *Eurozine*, 15 March. Available at <www.eurozine. com/pdf/2002-03-15-zizek-en.pdf> [last access on 2 October 2005)

Democratic Initiatives Fund and Kyiv International Institute of Sociology (2005) *Opinions and Views of Ukraine's population – April 2005*, press release on the pan-Ukrainian sociological survey (Kiev: Democratic Initiatives Fund and Kyiv International Institute of Sociology).

Department for Statistics and Sociology of the Republic of Moldova (2004) *Statistical Yearbook of the Republic of Moldova*, 2004 (Chisinau: Biblioteca DSS).

Derhachov Olexandr (2003) 'Vazhkyi khrest istorychnoi druzhby', *Dzerkalo tyzhnia*, 15–21. November.

Derrida, Jacques (1973) *Différance, Speech and Phenomena and Other Essays on Husserl's Theory of Signs*, trans. David B. Allison (Evanston, Illinois: Northwestern University Press).

Dima, Nicholas (2001) *Moldova and the Transdnestr Republic: Russia's Geopolitics toward the Balkans* (Boulder, East European Monographs).

DIW (Deutsches Institut für Wirtschaftsforschung) (2001) 'Wohlstandsgefälle in der EU-27 und Konsequenzen für die EU-Strukturpolitik', *DIW-Wochenbericht*, 68 (36): 562–66.

Dürrschmidt, Jörg (2002) '"They're worse off than us" – The Social Construction of European Space and Boundaries in the German/Polish Twin-City Guben-Gubin', *Identities: Global Studies in Culture and Power*, 9: 123–50.

Dutu, Mircea 2004) 'Legal Implications of the "Bystroe" Danube–Black Sea Canal Project'. *Revista Romana de Drept al Mediului*, 2(4): 1–19.

Dyckhoff, Claus (2004) 'Grenzübergreifende Zusammenarbeit in der Raumentwicklung – Erfahrungen aus der Sicht der Region Berlin-Brandenburg', in Michael Stoll (ed.), *Strukturwandel in Ostdeutschland und Westpolen*, Arbeitsmaterial 311 (Hannover: Akademie für Raumforschung und Landesplanung).

EC Treaty (2002) 'Consolidated Version of the Treaty on European Union', *Official Journal of the European Communities*, C325/6-181. Available at <europa.eu.int/ eur-lex/lex/en/treaties/dat/12002M/pdf/12002M_EN.pdf> [last access on 6 October 2005].

Éger, György (2000) *Regionalizmus, határok és kisebbségek Kelet-Közép-Európában* (Budapest, Osiris).

Ehrhart, Hans-Georg (1997) 'Die Ukraine und der Westen', *Europäische Rundschau*, (2): 49–59

Emerson, Michael (2001) *The Elephant and the Bear: The European Union, Russia and their Near Abroads*, (Brussels: Centre for European Policy Studies).

Erdősi, Ferenc and József Tóth (eds) (1988) *A sajátos helyzetű térségek terület- és településfejlesztési problémái*, Essays of the Symposium in Szombathely, 4–5 November 1986 (Pécs: Centre for Regional Studies of the Hungarian Academy of Science).

Euregio Karelia (2000) *Our Common Border 2001-2006* (Oulu: TACIS project).

European Parliament, Committee on Foreign Affairs, Human Rights, Common Security and Defence Policy (2003) Report on 'Wider Europe – Neighbourhood: A New Framework for Relations with our Eastern and Southern Neighbours', COM(2003) 104 – 2003/2018(INI), Available at <www2.europarl.eu.int/registre/

seance_pleniere/textes_deposes/rapports/2003/0378/P5_A(2003)0378_EN.doc> [last access on 7 October 2005].

European Spatial Planning Observation Network (2004) *Particular Effects of Enlargement of the EU and Beyond on the Polycentric Spatial Tissue with Special Attention on the Discontinuities and Barriers, Third Interim Report, Part I*, ESPON action 1.1.3 (Stockholm: The Royal Institute of Technology, Department of Infrastructure). Available at <www.espon.lu/online/documentation/projects/thematic/2057/1.1.1._part1.pdf> [last access on 11 October 2005].

Eskelinen, Heikki (2000) 'Cooperation across the Line of Exclusion: The 1990s Experience at the Finnish-Russian Border', *European Research in Regional Science (Borders, Regions and People)*, 10: 137–50.

EuroMed Civil Forum (2003) *Amending the EuroMed Civil Forum, Strengthening the EuroMed Civil Society Cooperation in the Barcelona Process*, Barcelona. Available at <www.euromedrights.net/barcelona-process/civil_society> [last access on 17 September 2003].

Falah, Ghazi and David Newman (1995) 'The Spatial Manifestation of Threat – Israelis and Palestinians Seek a Good Border', *Political Geography*, 14(8): 689–706.

Favell, Adrian and Randall Hansen (2002) 'Markets against Politics: Migration, EU Enlargement and the Idea of Europe', *Journal of Ethnic and Migration Studies*, 28 (4): 581–601.

Florida, Richard (2002) *The Rise of the Creative Class: And How It Is Transforming Work, Leisure, Community and Everyday Life* (New York: Basic Books).

Forsberg, Tuomas (1995) 'Karelia', in Tuomas Forsberg (ed.), *Contested Territory: Border Disputes at the Edge of the Former Soviet Empire* (Aldershot: Edward Elgar).

Foucher, Michel (1999) 'Europa and its Long-lasting Variable Geography', in Eberhard Bort and Russell Keat (eds), *The Boundaries of Understanding. Essays in Honour of Malcolm Anderson* (Edinburgh, The University of Edinburgh International Social Science Institute).

Franz, Peter (2004) 'Innovative Milieus in ostdeutschen Stadtregionen: "sticky places" der kreativen Klasse?', in Ulf Matthiesen (ed.), *Stadtregion und Wissen. Analysen und Plädoyers für eine wissensbasierte Stadtpolitik* (Wiesbaden: VS Verlag für Sozialwissenschaften).

Gheorghiu, Valeriu and Oxana Gutu (2004) 'River Prut: A Softer Iron Curtain', in Ann Lewis (ed.), *The EU & Moldova: On a Fault-Line of Europe* (London: The Federal Trust for Education and Research).

Gheorghiu, Valeriu, Ion Jigau and Natalia Vladicescu (2002) *Consequences of Schengen Treaty Implementation on Moldova's Western Border* (Warsaw: Institute for Public Policy, Moldova, in cooperation with the Institute of Public Affairs). Available at <www.isp.org.pl/files/28645991302871780011118306711.pdf> [last access on 1 October 2005].

Giddens, Anthony (1984) *The Constitution of Society: Outline of the Theory of Structuration* (Cambridge: Polity Press).

Golikov Artur P. and Pavel Chernomaz (1997) 'Evroregion "Slobozhanshchyna" kak forma transgranichnogo sotrudnichestva sopredelnykh oblastey Ukrainy I Rossii', *Region*, 4: 52–4.

Golobics, Pál (1996) 'A határ menti térségek városainak szerepe az interregionális együttmüködésben Magyarországon', in Ágnes Pál and Gabriella Szónokyné Ancsin (eds), *Határon innen - határon túl* (Szeged: Department of Economic Geography, József Atila University).

Golobics, Pál and József Tóth (1999) 'A nemzetközi regionális együttmüködés és Magyarország térszerkezete', in József Tóth and Zoltán Wilhelm (eds), *Változó környezetünk. Tiszteletkötet Fodor István professzor úr 60. születésnapjára* (Pécs: Geographical Institute of the Faculty of Natural Sciences, Janus Pannonius University, and Transdanubian Research Institute of the Centre for Regional Studies of the Hungarian Academy of Science).

Gorzelak, Gzegorz (1998) *Regional and Local Potential for Transformation in Poland*, Regional and Local Studies 14 (Warsaw: European Institute for Regional and Local Development).

Gorzelak, Grzegorz, John Bachtler and Mariusz Kasprzyk (eds) (2004) *Wspólpraca transgraniczna w Unii Europejskiej – doświadczenia Polski i Niemiec* (Cross-Border Cooperation in the European Union: Experiences of Poland and Germany) (Warsaw: Scholar).

Gorzelak, Grzegorz, Bohdan Jałowiecki, Richard Woodward, Wojciech Dziemianowicz, Mikolaj Herbst, Wojciech Roszkowski and Tomasz Zarycki (1999) *Dynamics and Factors of Local Success in Poland* (Warsaw: Warsaw University and Case Foundation).

Grabbe, Heather (2000) 'The Sharp Edges of Europe: Extending Schengen Eastwards', *International Affairs* 76(3): 481–514. Available at <www.cer.org.uk/pdf/grabbe_sharp_edges.pdf> [last access on 1 October 2005].

Guterres, Antonio (2001) *The European Treaties Revisited: What Role for Europe in the Globalised World?*, speech delivered at a conference at the Walter Hallstein-Institute for European Constitutional Law, Humboldt University, Berlin, 7 May.

Guz-Vetter, Marzenna (2002) *Polsko-niemieckie pogranicze. Szanse i zagrożenia w perspektywie przystąpienia Polski do Unii Europejskiej* (Warsaw: Instytut Spraw Publicznych).

Hajdú, Zoltán (1995a) 'Political Restructuring of East-Central Europe: Hungary as an Example', in Markku Tykkylainen (ed.), *Local and Regional Development during the 1990's Transition in Eastern Europe* (Aldershot: Avebury).

——— (1995b) 'A magyar államtér változásainak történeti és politikai földrajzi', *Tér és Társadalom*, 9 (3-4): 111–32.

Hargitai, Árpádné, Gabriella Izikné Hedril and Tibor Palánkai (eds) (1999) *Európa Kislexikon – az Európai Unió és magyarország* (Budapest: Aula).

Haukkala, Hiski (2001a) 'Succeeding Without Success? The Northern Dimension of the European Union', *Northern Dimensions: Yearbook 2001* (Helsinki: The Finnish Institute of International Affairs).

———— (2001b) *Two Reluctant Regionalizers? The European Union and Russia in Europe's North*, UPI Working papers, No. 32 (Helsinki: The Finnish Institute of Foreign Affairs).

Herd, Graeme P. (2001) 'Russian Systemic Transformation and Its Impact on Russo-Baltic Relations', in Pertti Joenniemi and Jevgenia Viktorova (eds), *Regional Dimensions of Security in Border Areas of Northern and Eastern Europe* (Tartu: Peipsi Centre for Transboundary Cooperation): 121–32.

Hettne, Björn (1999) 'Globalization and the New Regionalism: The Second Great Transformation', in Björn Hettne, András Inotai and Osvaldo Sunkel (eds) *Globalism and The New Regionalism*, vol. 1, (Basingstoke: Macmillan): 1–24.

Horváth Gyula (1998) *Európai regionális politika* (European Regional Policy) (Budapest and Pécs: Dialóg Campus).

———— (ed.) (2000) *A régiók szerepe a bővülő Európai Unióban* (Pécs: Centre for Regional Studies of the Hungarian Academy of Science).

Houtum, Henk van (2003) 'Borders of Comfort: Ambivalences in Spatial Economic Bordering Processes in and by the European Union', *Regional and Federal Studies*, 12: 37–58.

Houtum, Henk van and Ton van Naerssen (2002) 'Bordering, Ordering and Othering', *Tijdschrift voor Economische en Sociale Geografie*, 93 (2): 125–36.

Ilies, Alexandru (2003) *Romania intre milenii: Frontiere, areale frontaliere si cooperare transfrontaliera* (Oradea: Editura Universitatii din Oradea).

———— (2004) *Romania: Euroregiuni* (Oradea: Editura Universitatii din Oradea).

Illés, Iván (1993)'A Kárpátok Eurorégió', *Valóság*. 6 (1993): 12–19.

———— (1997) 'A regionális együttműködés feltételei Közép-és Kelet-Európában' (The Conditions of Regional Cooperation in Central and Eastern Europe), *Tér és Társadalom*, 2. (1997): 17–28.

Inotai, András (1998) *On the Way – Hungary and the European Union: Selected Studies* (Budapest: Belvárosi Könyvkiadó and International Business School).

Institut für Stadtforschung und Strukturpolitik GmbH (2004) *Fortschritte und Hemmnisse beim Vollzug des Stadtumbaus Ost zur Unternehmensumfrage. Endbericht.* (Berlin: Institut für Stadtforschung).

Ionescu, Valentin (2000) 'De la un protocol semnat... la lumanare, la o remarcabila extensie universitara', *Viata Libera*, September.

Izikné Hedri, G (ed.) (1995) *Magyarország úton az Európai Unióba* (Budapest: Aula).

Jandl, Michael (2003) *Moldova Seeks Stability Amid Mass Migration* (Washington, DC: International Centre for Migration Policy Development). Available at <www.migrationinformation.org/Profiles/display.cfm?id=184> [last access on 1 October 2005].

Janning, Josef and Werner Weidenfeld (eds) (1993) *Europe in Global Change* (Gütersloh: Bertelsmann Foundation).

Jenkins, Richard (1996) *Social Identity* (London: Routledge).

Joenniemi, Pertti (1996) 'Interregional Cooperation and a New Regionalist Paradigm', in James Scott, James, Alan Sweedler, Paul Ganster and Wolf-Dieter Eberwein (eds.), *Borders Regions in Functional Transition: European and North American Perspectives*, Erkner by Berlin: Institut für Regionalentwicklung und Strukturplanung): 53–61.

———— (1999) *Bridging the Iron Curtain? Co-operation around the Baltic Rim*, Working Papers of the Copenhagen Peace Research Institute, 22/1999 (Copenhagen: Copenhagen Peace Research Institute) (Contact through <www.copri.dk>).

———— (2002) *Can Europe Be Told from the North? Tapping into the EU's Northern Dimension*, Working Papers of the Copenhagen Peace Research Institute, 12/2002 (Copenhagen: Copenhagen Peace Research Institute) (Contact through <www.copri.dk>).

Kabisch, Sigrun, Matthias Bernt and Andreas Peter (2004) *Stadtumbau unter Schrumpfungsbedingungen. Eine sozialwissenschaftliche Fallstudie* (Wiesbaden: VS Verlag für Sozialwissenschaften).

Karl, Helmut(2002) 'Die Kontrolle nationaler Regionalbeihilfen in der EU', *Raumforschung und Raumordnung*, 3-4: 209–218.

Kennard, Ann (2003) 'The Institutionalization of Borders in Central and Eastern Europe: A Means to What End?', in Eiki Berg and Henk van Houtum (eds), *Routing Borders Between Territories, Discourses and Practices* (London: Ashgate).

———— (2004) 'Cross-border governance at the future eastern edges of the EU: a regeneration project?', in Olivier Kramsch and Barbara Hooper (eds) *Cross-border governance in the European Union* (London: Routledge).

Khokhotva, Ivan (2003) 'Dead on Arrival', *Transitions Online,* 26 September. Available at <www.tol.cz/look/TOL/article_single.tpl?IdLanguage=1&IdPublication=4&NrIssue=50&NrSection=3&NrArticle=10724&ST1=body&ST_T1=tol&ST_AS1=1&ST_max=1> [last access on 5 October 2005].

King, Charles (2000a) *The Moldovans: Romania, Russia, and the Politics of Culture* (Stanford, Hoover Institution Press).

———— (2000b) 'Diaspora Politics Between Ukraine and Romania', in James Clem and Nancy Popson (eds), *Ukraine and Its Western Neighbors* (Washington: East European Studies, Woodrow Wilson Center).

Király, Béla K. and Ignác Romsics (eds.) (1998) *Geopolitics in the Danube region: Hungarian Reconciliation Efforts, 1848-1998* (Budapest: Central European University Press).

Kiryukhin, Aleksey (2000) 'Territorialnaia struktura Evroregiona "Slobozhanshchyna"', *Biznes-inform*, 6: 48-50.

Kissinger, Henry (1994) *Diplomacy* (New York: Simon and Schuster).

Klesmann, Martin and Magdalena Kursa (2003) 'Frankfurt will Zuzügler vom anderen Oderufer. Mieter aus Polen sollen Wohnungsleerstand beheben', *Berliner Zeitung*, 20 (27 Mai 2003).

Klochkov, Yuriy and Aleksey Kiryukhin (2001) 'Prigranichny polyus rosta: perspektivy ozhivlenia torgovli i strudnichestva', *Vestnik torgovo-promyshlennoy palaty. Sovmestnyi vypusk Kharkovskoy I Belgorodskoy torgovo-promyshlennykh palat*, 10: 16.

Kocsis, Karoly and Eszter Kocsis-Hodosi (1998) *Hungarian Minorities in the Carpathian Basin: A Study in Ethnic Geography* (Toronto: Matthias Corvinus Publishing).

Kolossov Vladimir and Olga Vendina (2002) 'Rossiysko-ukrainskaia granitsa: sotsialnye gradienty, identichnosti i migratsionnye potoki (na primere Belgorodskoy i Kharkovskoy oblastey)', National Institute of International Security Problems (ed.) *Migratsia i pogranichnyi rezhim: Belarus, Moldova, Rossia i Ukraina*, (Kiev: National Institute of International Security Problems).

Kolossov Vladimir and Aleksey Kiryukhin (2001) 'Prigranichnoe sotrudnichestvo v rossiysko-ukrainskikh otnosheniakh', *Politia*, 1 (19): 141–65.

Kononenko, Vadim (2004) *What's New About Today's EU-Russia Border*, UPI working papers 50 (Helsinki: Ulkopoliittinen Instituutti/The Finnish Institute of International Affairs).

Kramsch, Oliver (2003) 'Re-imagining the "Scalar Fix" of Transborder Governance: the Case of the Maas-Rhein Euregio', in Eiki Berg and Henk van Houtum (eds) *Routing Borders between Territories, Discourses and Practices* (Aldershot: Ashgate).

——— (2004) 'Towards a Mediterranean Scale of Governance? 21st-Century Urban Networks across the "Inner Sea"', in Olivier Kramsch and Barbara Hooper (eds) *Cross-Border Governance in the European Union* (London: Routledge).

Kramsch, Olivier and Barbara Hooper (eds) (2004) *Cross-Border Governance in the European Union* (London: Routledge).

Kramsch, Olivier, Roos Pijpers, Roald Plug, and Henk van Houtum (2004) *Research on the Policy of the European Commission towards the Re-bordering of the European Union*, research report prepared for EXLINEA, European Commission's Fifth Framework Programme (Nijmegen: Radboud University). Available at <www.ru.nl/gap/papers/gapwp04-07.pdf> [last access on 2 October 2005].

Krätke, Stefan (1996) 'Where East Meets West: the German-Polish Border Region in Transformation', *European Planning Studies*, 4 (6): 647–69.

——— (1998) 'Problems of Cross-Border Regional Integration: The Case of the German-Polish Border Area', *European Urban and Regional Studies*, 5 (3): 249–62.

——— (1999) 'Regional Integration or Fragmentation? The German-Polish Border Region in a New Europe', *Regional Studies*, 33 (7): 631–41.

——— (2002) 'Cross-Border Cooperation and the Regional Development in the German–Polish Border Area', in Markus Perkmann and Ngai-Ling Sum (eds), *Globalization, Regionalization, and Cross-Border Regions* (Houndsmill: Palgrave Macmillan).

——— (2004) 'Perspektiven der EU-Osterweiterung für regionale Wirtschaftszentren und Grenzregionen', in Michael Stoll (ed.), *Strukturwandel in Ostdeutschland und Westpolen*, Arbeitsmaterial 311 (Hannover: Akademie für Raumforschung und Landesplanung).

Krätke, Stefan and Renate Borst (2004) *EU-Osterweiterung als Chance. Perspektiven für Metropolräume und Grenzgebiete am Beispiel Berlin-Brandenburg*, Beiträge zur europäischen Stadt- und Regionalforschung 1 (Münster: LIT-Verlag).

Krätke, Stefan and Christoph Scheuplein (2001) *Produktionscluster in Ostdeutschland. Methoden der Identifizierung und Analyse* (Berlin, Verlag VSA).

Krok, Katarzyna (2004) 'Polsko-ukraińska wspólpraca transgraniczna a polityka Unii Europejskiej', in Wojciech Bieńkowski, Jerzy Grabowiecki and Henryk Wnorowski (eds), *Rozszerzenie Unii Europejskiej na Wschód a Rozwój Współpracy Transgranicznej* (Białystok, Uniwersytet w Białymstoku-Wydział Ekonomiczny).

Kruglashov Anatoliy (2004) 'Etnopolitychna harmonizatsiya: chy pid sylu tse zavdannia novoutvorenum Evroregionam?", *Evroregiony: potentsial mizhetnichnoyi harmonizatsiyi*, (Chernivtsi: Bukrek).

Kühn, Manfred (2004) 'Wissenschaft Stadt. FuE-basierte Siedlungsentwicklung in deutschen Stadtregionen', in Ulf Matthiesen (ed.), *Stadtregion und Wissen. Analysen und Plädoyers für eine wissensbasierte Stadtpolitik* (Wiesbaden, VS Verlag für Sozialwissenschaften).

Laffan, Brigid, Rory O'Donnell and Michael Smith (2000) *Europe's Experimental Union: Rethinking Integration* (London: Routledge).

Laidi, Zaki (1998) *A World without Meaning. The Crisis of Meaning in International Politics*, (London: Routledge).

Light, Margot, Stephen White, Stephen and John Löwenhardt (2000) 'A Wider Europe: The View from Moscow and Kyiv', *International Affairs*, 76, 1: 77–88.

Liikanen, Ilkka (2004) 'Euregio Karelia: A Model for Cross-Border Cooperation with Russia?', *Journal for Foreign and Security Policy*, 1 (3). Available at <www.iiss.org/rrpfreepdfs.php?scID=65> [last access on 7 October 2005].

Lippert, Barbara (2001) *Border Issues and Visa Regulations: Political, Economic, and Social Implications* (Gütersloh: Bertelsmann Foundation). Available at <www.euintegration.net/data/doc_publications/213/Lippert.pdf> [last access on 1 October 2005].

Lipponen, Paava (1997) *The European Union Needs a Policy for the Northern Dimension*, speech presented at the Barents Region Today conference, Rovaniemi, Finland, 15 September 1997.

——— (2002) Speech presented at the Conference on Partnership and Growth in the Baltic Sea Region, Copenhagen, 13 October 2002. Available at <http://www.valtioneuvosto.fi/vn/liston/base.lsp?r=24372&k=en&old=1300> [last access on 22 November 2005]

Lobjakas, Ahto (2004) 'European Commission Unveils Details of "New Neighborhood" Strategy', Radio Free Europe/Radio Liberty. Available at <rferl.com/featuresarticle/2004/05/077d3a6b-3883-4119-972e-cc470a9ff6b6.html> [last access on 1 October 2005].

Loughlin, John (2001) *Subnational Democracy in the European Union: Challenges and Opportunities* (Oxford: Oxford University Press).

Ludvig, Zsuzsa (2003) *Hungarian-Ukrainian Cross-border Cooperation with Special Regard to Carpathian Euroregion and Economic Relations* (Warsaw: Batory Foundation). Available at <www.batory.org.pl/doc/l1.pdf > [last access on 1 October 2005].

Luhmann, Niklas (1985) *Soziale Systeme, Grundriss einer Allgemeinen Theorie,* second edition (Frankfurt am Main, Suhrkamp).

McGrew, Andrew (2000) 'Power Shift: From National Government to Global Governance?' in David Held (ed), *A Globalizing World? Culture, Economics, Politics* (London and New York: Routledge): 127–67.

Malynovska, Olena (2002) *EU Enlarged, Schengen Implemented — What Next? Political Perspectives for Ukraine* (Warsaw: International Centre for Policy Studies, Ukraine, in cooperation with the Institute of Public Affairs). Available at <www.isp.org.pl/files/4755113720077644001118216408.pdf> [last access on 1 October 2005].

Matthiesen, Ulf and Hans-Joachim Bürkner (2001) 'Antagonistic Structures in Border Areas. Local Milieux and Local Politics in the Polish-German Twin City Gubin/Guben', *GeoJournal,* 54: 43–50.

———— (2002) *Grenzmilieus im potentiellen Verflechtungsraum von Polen mit Deutschland,* IRS Working Papers, 02/2002, (Erkner: Institut für Regionalentwicklung und Strukturplanung) Available at <www.irs-net.de/download/grenzmilieus.pdf) [last access on 11 October 2005].

———— (2004) 'Wissensmilieus – Zur sozialen Konstruktion und analytischen Rekonstruktion eines neuen Sozialraum-Typus', in Ulf Matthiesen (ed.), *Stadtregion und Wissen. Analysen und Plädoyers für eine wissensbasierte Stadtpolitik* (Wiesbaden: Leske & Budrich).

Matthiesen, Ulf and Jörg Dürrschmidt (2002) 'Everyday Milieux and Culture of Displacement. A Comparative Investigation into Space, Place and (Non)Attachment within the German-Polish Twin City Guben/Gubin', *Canadian Journal of Urban Research,* 11: 17–45.

Mazur, Agnieszka (2002) 'The EU's Eastern Dimension – Echoes of the Northern Dimension?', in Finnish Institute of International Affairs (FIIA/UPI), *The New North of Europe,* Policy Memos of the Finnish Institute of International Affairs (UPI), (Helsinki: Finnish Institute of International Affairs): 59–62.

Meinhof, Ulrike H. (2000) *Living (with) Borders: Identity Discourses on East-West Borders in Europe* (Aldershot: Ashgate).

Ministry of Economy of Ukraine (2003) *Draft Law on Stimulation of Regional Development* (Kiev: Ministry of Economy of Ukraine).

Ministry of Economy of Ukraine (2004), *Draft National Strategy for Regional Development 2005–2015* (Kiev: Ukrainian Ministry of Economy).

Ministry of Foreign Affairs of Ukraine (2004) *EU-Ukraine Action Plan, 2005-2007* (Kiev: Ministry of Foreign Affairs of Ukraine).

Mittelman, James (1999) 'Rethinking the New Regionalism in the Context of Globalization', in Björn Hettne, András Inotai and Osvaldo Sunkel (eds.),

Globalism and the New Regionalism, vol. 1 (Basingstoke: Macmillan): 25–53.

Moisi, Petruta (2004) 'Huiduiti in Ucraina', *Viata Libera*. 14 October.

Murphy, Alexander (1996) 'The Sovereign State System as Political-Territorial Ideal: Historical and Contemporary Considerations', in Thomas J. Biersteker and Cynthia Weber (eds), *State Sovereignty as Social Construct* (Cambridge: Cambridge University Press, 1996).

National Institute for Statistics (2002) Population Data 2001 (Bucharest: NIS).

Negut, Silviu (1998) 'Les Euroregions', *Revue Roumaine de Géographie*, 42: 75–85.

Nemes, Nagy, József (1998) *Tér a társadalomkutatásban* (Budapest: Hilschler Rezső Szociálpolitikai Egyesület).

Neumann, Iver (1999) *Uses of the Other. 'The East' in European Identity Formation* (Minneapolis: University of Minnesota Press).

Newman, David (2003) 'Boundary Geopolitics: Towards a Theory of Territorial Lines?', in Eiki Berg and Henk van Houtum (eds), *Routing Borders between Territories, Discourses and Practices* (Aldershot: Ashgate).

O'Dowd, Liam (2002) 'The Changing Significance of European Borders', *Regional and Federal Studies*, 12 (4): 13–36.

——— (2003) 'The Changing Significance of European Borders', in James Anderson, Liam O'Dowd and Thomas M. Wilson (eds), *New Borders for a Changing Europe: Cross-Border Cooperation and Governance*, (London: Frank Cass).

O'Loughlin, John and Herman van der Wusten (eds) (1993) *The New Political Geography of Eastern Europe* (London: Belhaven Press).

O'Rourke Breffni (2003) 'EU: Prodi Sets out Vision of "Ring of Friends", Closer Ties with Neighbors', *Radio Free Europe / Radio Liberty.* Available at <rferl. org/nca/features/2003/01/31012003192138.asp > [last access on 5 October 2005].

Owen, David (1995) *Balkan Odyssey* (London: Victor Gollanz).

Pándi, Lajos (1995) *Köztes-Európa 1763-1993* (Budapest: Osiris-Századvég).

Pap, Norbert and Jószef Tóth (eds) (2002) *Európa politikai földrajz* (Pécs, Alexandra Könyvkiadó).

Patten, Chris (2001) *Common Strategy for the Mediterranean and Reinvigorating the Barcelona Process*, speech delivered at the European Parliament, joint debate, 31 January 2001 in Brussels, SPEECH/01/49. Available at <europa.eu.int/comm/external_relations/news/patten/speech_01_49.htm> [last access on 22 November 2005].

Perkmann, Markus and Ngai-Ling Sum (2000) 'Introduction', in Markus Perkmann and Ngai-Ling Sum (eds), *Globalization, Regionalization, and Cross-Border Regions* (Houndsmill: Palgrave Macmillan).

——— (2002) 'Euroregions: Institutional Entrepreneurship in the European Union', in Marko Perkmann and Ngai-Ling Sum (eds), *Globalization, Regionalization and Cross-Border Regions* (Houndsmills: Palgrave).

Pérez Diaz, Victor (1994) *The Challenge of the European Public Sphere*, ASP Research Paper, 4b/1994. Available at <www.asp-research.com/publications. asp?tipo=2> [last access on 22 November 2005].

Philip Morris Institute (1998) *Is the Barcelona Process Working? EU Policy in the Eastern Mediterranean* (Brussels: Philip Morris Institute for Public Policy Research).

Piipponen, Risto (2003) 'EU:n ulkosuhteiden kokonaisuus', in Pauli Järvenpää, Kirsti Kauppi, Olli Kivinen, Hanna Ojanen, Risto Piipponen, Olli Rehn and Antii Sierla, *Suomen paikka maailmassa? EU:n ulko- ja turvallisuuspolitiikka* (Helsinki: Edita Prima Oy).

Piskozub, Andrzej (1987) *Dziedzictwo polskiej przestrzeni* (Wrocław: Ossolineum).

Popa, Razvan (2004) 'Securizarea Frontierelor – ori 800 milioane, ori 1 miliard de euro', *Adevarul*, 1 October.

Prodi, Romano (2001) *Solidarity, the Foundation on Which Europe Stands*, speech delivered at the opening session of the European Forum on Economic and Social Cohesion, Brussels, 21 May 2001, SPEECH/01/236. Available at <europa.eu.int/rapid/pressReleasesAction.do?reference=SPEECH/01/ 236&format=HTML&aged=0&language=EN&guiLanguage=en> [last access on 22 November 2005].

Prohnitsky, Valeriu (2002) 'Moldova-Ukraine-Romania: A Regional Portrayal of Economy and Trade', *South-East Europe Review for Labour and Social Affairs,* 5(2) (2002): 35-48. Available at <www.boeckler.de/pdf/South-East_Europe_ Review-2002-02-s35.pdf> [last access on 1 October 2005].

Rechnitzer, János (1999a) 'Országhatár menti együttműködések, mint a területfejlesztés új stratégiai irányai', in Gabriella Szónokyné Ancsin (ed.), *Borders and Regions* (Szeged: Department of Economic and Social Geography, Faculty of Natural Sciences, University of Szeged).

—— (1999b) 'Határ menti együttműködések Európában és Magyarországon. – Elválaszt és összeköt – a határ. Társadalmi-gazdasági változások az osztrák–magyar határ menti térségekben', in Márta Nárai and János Rechnitzer (eds), *Elválaszt és összeköt – a határ* (Pécs–Győr: Centre for Regional Studies of Hungarian Academy of Science).

Reut, Oleg (2000) *Republic of Karelia: A Double Asymmetry of North-Eastern Dimensionalism*, Working Papers of the Copenhagen Peace Research Institute, 12/2000 (Copenhagen: Copenhagen Peace Research Institute) (Contact through <www.copri.dk>).

—— (2002) 'Euroregion Karelia: More Feeling than Substance', Moscow, Moscow School of Political Studies Internet Project.

Ring, Éva et al. (ed.) (1986) *Helyünk Európában. Nézetek és koncepciók a 20. századi*, vol. 1 and 2 (Budapest: Magvető Könyvkiadó).

Romsics, Ignac (1996) *Helyünk és sorsunk a Duna-medencében* (Budapest: Osiris Kiadó).

Ruttkay Éva (1995) 'Határok, határmentiség, regionális politika', *Comitatus*, 12: 23–35.

Sallai, János (2003) *Kishatárforgalom, kelet-magyarországi határkapcsolat* (Bilateral Border Crossing, Cross-Border Relations in Eastern Hungary) (Budapest: Rendőrtiszti Főiskola).

Sassen, Saskia (1988) *The Mobility of Labour and Capital: A Study in International Investment and Labor Flow* (Cambridge: Cambridge University Press).

——— (2002) 'Is this the Way to Go? Handling Immigration in a Global Era', *Eurozine*, 17 September. Available at <www.eurozine.com/pdf/2002-09-17-sassen-en.pdf> [last access on 1 October 2005].

Scott, James W. (1996) 'Dutch–German Euroregions: A Model for Transboundary Cooperation', in James Wesley Scott, Alan Sweedler, Paul Ganster and Wolf-Dieter Eberwein (eds), *Border Regions in Functional Transition: European and North American Perspectives* (Berlin, Institut für Regionalentwicklung und Sozialforschung).

——— (1997) 'A határ menti együttműködés nemzetközi rendszerei', *Tér és Társadalom*, 3: 117–31.

——— (1998) 'Planning Co-operation and Transboundary Regionalism: Implementing European Border Region Policies in the German-Polish Context', *Environment and Planning C: Government and Policy*, 16 (5): 605–624.

——— (1999) 'European and North American Contexts for Cross-Border Regionalism', *Regional Studies*, 33 (7): 605–617.

——— (2000) *Transnational Regionalism, Strategic Geopolitics and European Integration: The Case of the Baltic Sea Region*, paper presented at the Border Research Network Second Workshop in Aabenraa, Denmark, 'How Do Borders Matter in Contemporary Europe', 25–27 May.

——— (2002) 'A Networked Space of Meaning? Spatial Politics as Geostrategies of European Integration', *Space and Polity*, 6(2): 147–67.

——— (2004) 'The Northern Dimension: "Multiscalar" Regionalism in an Enlarging European Union?', in Olivier Kramsch and Barbara Hooper (eds), *Cross-Border Governance in the European Union* (London: Routledge, 2004).

——— (2005) 'The EU and "Wider Europe": Toward an Alternative Geopolitics of Regional Cooperation?', *Geopolitics*, 10 (3): 429–54.

——— (2005) 'Transnational Regionalism, Strategic Geopolitics and European Integration: The Case of the Baltic Sea Region', in Nicol Heather and Ian Townsend Gault (eds), *Holding the Line. Borders in a Global World* (Vancouver: University of British Columbia Press).

Scott, James W. and Kimberly Collins (1997) 'Inducing Transboundary Regionalism in Asymmetric Situation: The Case of the German–Polish Border', *Journal of Borderlands Studies*, 12 (1–2): 97–121.

Selke, Welf (2004) 'Deutsch-Polnisches Haus: Auf dem Wege zu einer grenzüberschreitenden Wirtschaftsregion in Mitteleuropa?', in Michael Stoll (ed.), *Strukturwandel in Ostdeutschland und Westpolen*, Arbeitsmaterial 311 (Hannover: Akademie für Raumforschung und Landesplanung).

Serebrian, Oleg (2004) '"Good Brothers", Bad Neighbours: Romanian/Moldovan Relations', in Ann Lewis (ed.), *The EU & Moldova: On a Fault-Line of Europe* (London: The Federal Trust for Education and Research).

Sergounin, Alexander (2001) 'Russia and Transborder Security Interests in Northern Europe: Defining a Co-operative Agenda', in Pertti Joenniemi and Jevgenia Viktorova (eds), *Regional Dimensions of Security in Border Areas of Northern and Eastern Europe* (Tartu: Peipsi Centre for Transboundary Cooperation).

Shafir, Michael (1985) *Romania: Politics, Economics and Society* (London, Frances Pinter).

Shlyamin, Valery (2004) *Russia in the Northern Dimension*. (Oulu: Euregio Karelia).

Sibley, David (1995) *Geographies of Exclusion: Society and Difference in the West* (London: Routledge).

―――― (2001) 'The Binary City', *Urban Studies*, 38 (2): 239–50.

Sík, Endre (2002) 'Külföldiek Magyarországon és a velük kapcsolatos nézetek a helyi önkormányzatokban', in Gábor Sisák (ed.) *Nemzeti és etnikai kisebbségek Magyarországon a 20. Század végén* (Budapest, Osiris-MTA Kisebbségkutató Műhely).

Siposné Kecskeméthy, Klára and Miklós Nagy (1995) 'A magyar katonaföldrajz alapkérdésének változása és vizsgálati mutatói', *Földrajzi Értesítő*, XLIV (1-2): 71–89.

Sjursen, Helene (2003) *Security and Defence*, ARENA Working Paper Series 10/03 (Oslo: ARENA Centre for European Studies, University of Oslo).

Skvortova, Alla (2001) 'Moldova and the EU: Direct Neighbourhood and Security Issues', in Iris Kempe (ed.), *Beyond EU Enlargement*, vol. 1, *The Agenda of Direct Neighbourhood for Eastern Europe* (Gütersloh: Bertelsmann Foundation.

―――― (2002) 'The Cultural and Social Makeup of Moldova: A Bipolar or Dispersed Society', in Pal Kolsto (ed.), *National Integration and Violent Conflict in Post-Soviet Societies: The Cases of Estonia and Moldova* (Lanham: Rowman & Littlefield).

Smętkowski, Maciej (2002) 'Rola położenia przygranicznego w rozwoju lokalnym – przykład miasta Przemyśla', *Kwartalnik Studia Regionalne i Lokalne*, 2-3: 141–55.

―――― (2005) 'Ewaluacja Programu Współpracy Przygranicznej Polska-Niemcy – ujęcie sektorowe: transport', in Grzegorz Gorzelak, John Bachtler and Mariusz Kasprzyk (eds) *Współpraca transgraniczna Unii Europejskiej: doświadczenia polsko-niemieckie*, Wydawnictwo (Warsaw: Naukowe Scholar).

Smith, Michael (1996) 'The European Union and a Changing Europe: Establishing the Boundaries of Order', *Journal of Common Market Studies*, 34 (1): 5–28.

State Statistics Committee (2001) Ukrainian Census Data 2000 (Kiev: SSC).

State Statistics Committee (2004) *Ukraine in Figures, 2003* (Kiev: State Statistics Committee).

Steering Group of NDEP (2002) *The Northern Dimension Environmental Partnership: Pledging Conference for the NDEP Support Fund*, 9 July 2002. Available at <europa.eu.int/comm/external_relations/north_dim/ndep/> [last access on 22 November 2005].

Stein, Rolf (1997) *Restructuring of a Traditional Sector in the New Europe and Regional Development of the Polish-German Border Area - Integration or Fragmentation? Some Evidence from the Wood Products and Furniture Industry. Frankfurt/Oder*, Analysen und Diskussionsbeiträge (Frankfurt/Oder: Lehrstuhl für Wirtschafts- und Sozialgeographie, 5).

Story, Jonathan (ed.) (1993) *The New Europe: Politics, Government and Economy since 1945* (Oxford: Blackwell).

Stryjakiewicz, Tadeusz (2004) 'Industrial Change in Western Poland against the Background of the Transformation of the National Economy', in Michael Stoll (ed.), *Strukturwandel in Ostdeutschland und Westpolen*, Arbeitsmaterial 311 (Hannover: Akademie für Raumforschung und Landesplanung).

Sum, Ngai-Ling (2002) 'Globalization, Regionalization and Cross-Border Modes of Growth in East Asia: the (Re-)Constitution of "Time-Space Governance"', in Markus Perkmann and Ngai-Ling Sum (eds), *Globalization, Regionalization and Cross-Border Regions* (Basingstoke: Palgrave Macmillan): 50–76.

Tatur, Melanie (2004) 'Introduction: Conceptualising the Analysis of 'making Regions' in post-Socialist Europe', in Melanie Tatur (ed.), *The Making of Regions in Post-Socialist Europe – the Impact of Culture, Economic Structure and Institutions. Case Studies from Poland, Hungary, Romania and Ukraine*, Vol. I (Wiesbaden: VS Verlag für Sozialwissenschaften).

Telo, Mario (ed.) (2001) *European Union and the New Regionalism. Regional Actors and Global Governance in a Post-Hegemonic Era* (Aldershot: Ashgate).

Terra Studio (2000) *Magyar–román határ menti térségek fejlesztési koncepciója és programja* (Budapest, Terra Studio Kft).

Terva, Jyrki (1999) *The Northern Dimension: Changing Borders to Frontiers or Escaping the Shadows of the Past?*, unpublished manuscript, University of Tampere.

Tóth, József (1997) 'Régiók a Kárpát-medencében. – Európa politikai földrajza', in Norbert Pap and József Tóth (eds), *Európa politikai földrajza* (Pécs: Department of Social Geography and Urban Studies, Faculty of Natural Sciences of the Janus Pannonius University).

Tunander, Ola, Pavel Baev and Victoria-Ingrid Einagel (eds) (1997) *Geopolitics in Post-Wall Europe: Security, Territorry and Identity* (London: SAGE).

Urry John (2000) *The Global Media and Cosmopolitanism*, (Lancaster, Department of Sociology, Lancaster University). Available at <www.lancs.ac.uk/fss/sociology/papers/urry-global-media.pdf> [last access on 2 October 2005].

Urwin, Derek W. (1999) *A közös Európa. Az európai integráció 1945-től napjainkig* (Budapest: Corvina).

Verheugen, Günter (2003) *The Eastern Dimension*, speech delivered at Vilnius, Lithuania, 25 April 2003. Available at <www.eudel.lt/en/special_features/

Files/03-Verhegen%20speech%20in%20Seimas%20on%20Eastern%20Dim
ension.doc?client=0b8093294c51dd1f486ce7dd5ac288f3> [last access on 29
September 2005].

Virtanen, Petri (2004) 'Euroregions in Changing Europe: Euregio Karelia and
Euroregion Pomerania as Examples', in Olivier Kramsch and Barbara Hooper
(eds), *Cross-Border Governance in the European Union* (London: Routledge).

Vorob'eva, L. M. (2004) 'Moldavia i ES: Formirovanie bazy sotrudnichestva', in
Evgenii M. Kozhokin (ed.), *Moldavia: Sovremennye tendentsii razvitia* (Moscow:
The Russian Political Encyclopaedia).

Wallace, Claire (2002) 'Opening and Closing Borders: Migration and Mobility in
East-Central Europe', *Journal of Ethnic and Migration Studies*, 28 (4): 603–625.

Wallace, William (2003) *Looking after the Neighbourhood: Responsibilities for the
EU-25*, Policy Papers of Notre Europe–Groupement des Études et de Recherches,
4. Available at <www.notre-europe.asso.fr/IMG/pdf/Policypaper4.pdf> [last
access on 29 September 2005].

Walzer, Michael (1983) *Spheres of Justice* (New York: Basic Books).

Weidenfeld, Werner and Franz-Lothar Altmann (ed.) (1995) *Central and Eastern
Europe on the Way into the European Union: Problems and Prospects of
Integration* (Gütersloh: Bertelsmann Foundation).

Weislo, Paulina (2004) *Deutsch-Polnisches Haus: Innovationsimpulse durch ein
grenzüberschreitendes Städtenetz*, diploma thesis presented at the Freie Universität
Berlin (Hamburg: Diplomica GmbH).

Wolchik, Sharon L. (1991) *Czechoslovakia in Transition: Politics, Economics, and
Society* (London: Pinter).

Zimmer, Kerstin (2004) 'The Captured Region: Actors and Institutions in the
Ukrainian Donbas', in Melanie Tatur (ed.) *The Making of Regions in Post-Socialist
Europe – the Impact of Culture, Economic Structure and Institutions: Case Studies
from Poland, Hungary, Romania and the Ukraine*, vol. II (Wiesbaden: VS Verlag
für Sozialwissenschaften).

Zizek, Slavoj (1998) *Pleidooi voor intolerantie* (Amsterdam: Boom).

Index

Locators shown in italics refer to figures, maps and tables.

Action Plan for the Northern Dimension (EU, 2000), 118
Action Plan on Justice and Home Affairs in Ukraine (EU, 2003), 101
Agreement on the Cooperation of the Border Regions [Russia/Ukraine, 1995], 97
agriculture, European
    SAPARD, 120
    statistics of growth, *92*
Algeria
    geopolitical relations with EU, 27-32
Alpine-Adriatic Working Community (1978), 70, 154
Antall, J., 74

Baltics, The
    role of EU in, 23-4
Baranyi, B., 163
Basescu, C., 148
Bauman, Z., 55
Békés County Business Zone, 159
Benhabib, S., 57
Bihar-Bihor euroregion, *152,* 153, *153,* 157-9, *158*
Bihar Business Zone, 159
borders and borderlands
    management motives, 54-5
    perceptions and perspectives concerning, 3-6
    socioeconomic challenges, 187-9, *189,* 198-9, 208-13, *208*
    see also conflicts, borderland; cooperation, cross-border; entrepreneurship, cross-border; euroregions; trade, cross-border; traffic, cross-border; travel, cross-border
    see also countries eg Germany; Hungary Moldova; Poland; Romania; Ukraine
Brandenberg, Eastern
    socioeconomic challenges, 209-11, 212-13

Brandt, W., 197
Bulgaria
    macroeconomic growth, *92*
Bürkner, H-J., 195
businesses
    role in inter-regional cooperation, 159
Bystroe Canal, 45-6

CBC (cross-border cooperation) *see* cooperation, cross-border
Carpathian Basin
    geopolitical fragmentation, 150-51
    see also country name eg Hungary; Romania
Carpathian Euroregion, *39,* 85, *84, 86, 87,* 154-7, 167
Catellani, N., 26
Central region (Ukraine), *84,* 85, *86, 87*
cities
    role in inter-regional cooperation, 159-60
COMECON (Council for Mutual Economic Assistance), 69-70
Common Foreign and Security Policy (CFSP) (EU), 117
Common Strategy on Russia (EU, 1999), 119
conflicts, borderland
    involving Moldova/Romania/Ukraine, 44-8
Consortium of Near-Border Ukrainian and Russian Universities, 97
cooperation, cross-border
    EU policy threats, 40-41
    local attitudes towards, 165-9, *166, 167, 168, 169*
    outcomes of EU policies, 47-50
    Ukrainian potential for, 87-91
    value of 'Northern Dimension' perspective, 118

see also trade, cross-border; traffic, cross-border
see also forms and countries eg euroregions; Poland; Russia; Ukraine
Council for Mutual Economic Assistance (COMECON), 69-70
Council of the Border Regions of Russia and Ukraine, 97
Council of the Lower Danube Euroregion, 143
Crongberg, T., 128-9

Danube-Körös-Maros-Tisza (DKMT) euroregion, 153, *153*, 154-7, 167
Derhachov, O., 105
doctrines, geopolitical
  emergence concerning 'Wider Europe' concept, 19-23
Donetsky region (Ukraine), 83, *84, 86*

East Brandenberg
  socioeconomic challenges, 209-11, 212-13
Eastern region (Ukraine), 83, *84, 85, 86, 87*
Economic and Social Committee (ESC) (EU), 31
Egypt
  geopolitical relations with EU, 27-32
EMD (Euro-Mediterranean Partnership) strategy, 28-32
employment
  experience in German-Polish borderlands, 209-12
  trends and policies in Hungary, 172-6, *174, 175*
enlargement, European Union
  impact on Modovan-Romanian relations, 135-7
  impact on Russia, 100-104
  impact on Ukraine, 81-3, 87-93, *92*, 100-104
ENP (European Neighbourhood Policy), 18, 22, 82
ENPI (European Neighbourhood and Partnership Instrument), 90, 113, 114-15, 121-2
entrepreneurship, cross-border

barriers in Polish-Ukrainian borderlands, 188-9
see also trade, cross-border
ESC (Economic and Social Committee) (EU), 31
EU BORDER IDENTITIES (research project), 4
Euregio Karelia, 124-7, *125*
EuroMed Civil Forum, 30-31, 32
Euro-Mediterranean Partnership (EMD) strategy, 28-32
Europe
  migration policies, 53-4, 55-8
Europe, Eastern
  development of euroregions, 37-40, *39*
  EU policy threats, 40-41
European Neighbourhood and Partnership Instrument (ENPI), 90, 113, 114-15, 121-2
European Neighbourhood Policy (ENP), 18, 22, 82
European Neighbourhood Strategy Paper (EU, 2004), 122
European Union (EU)
  assistance in Polish borderlands, 187-9, *189*
  geopolitical relations with Mediterranean states, 27-32
  outcomes of borderland policies, 47-50
  policy threats to eastern Europe, 40-41
  role in Baltic region, 23-4
  see also cooperation, cross-border; enlargement, European Union; euruoregions
  see also country name eg Germany; Hungary; Poland
euroregions
  development in Eastern Europe, 37-40, *39*
  emergence and role in Hungary, 151-9, *152, 153, 156, 158*
  history and characteristics, 123-4, 151-2
  impact on Finland and Russia, 127-9
  involvement in Modovan-Romanian cooperation, 143-6, *144-5*
  local perceptions of locals, 167-9, *168, 169*
  location of, 202, *203*

socioeconomic challenges, 198-9, 208-
13, *208*
see also individual name eg
Bihar-Bihor; Euregio Karelia;
Slobozhanshchyna
EXLINEA (research project), 4, 5

Favell, A., 58
Finland
impact of euroregion development, 127-
9
see also Euregio Karelia
'Fortress Europe' (concept)
for EU migration policies, 56-58
see also 'Gated Communities'
Frankfurt-an-der-Oder
socioeconomic challenges, 209-11, 212-
13
funding, programmes of
cross-border control, 120-21

'Gated Communities' (concept)
for EU migration policies, 59-61
see also 'Fortress Europe'
GDP (Growth Domestic Product)
EU statistics, *92*
Geonana, M., 139
geopolitics (concept)
emergence of, 19-23
see also country or element eg Algeria;
borders and borderlands
Germany
cross-border cooperation with Poland,
199-204, *201*, 209-12, 213-15
geopolitical history of borderlands, 196-
8, *197*
location of euroregions, 202, *203*
socioeconomic challenges, 198-9, 209-
13
see also trade, cross-border
see also regions eg Brandenberg, East
Gorzela, G., 212
government, local
Polish-Ukrainian cooperation, 186-7
government, national
Romanian-Moldovan relations, 133-8
Growth Domestic Product (GDP)
EU statistics, *92*
Guterres, A., 21

Hajdú-Bihar-Bihor euroregion, *152,* 153,
153, 158-9, *158*
Hansen, R., 58
Hettne, B., 20-21
Hombach, B., 142
Hungarian Democratic Forum, 73
Hungary
attitudes towards cross-border
cooperation, 165-9, *166, 167, 168,
169*
emergence and role of euroregions, 151-
9, *152, 153, 156, 158*
EU membership, 77-8
geopolitical history of borderlands, 68-9
geopolitics, 66-77, *67*
socioeconomic trends and policies, *92,*
172-6, *173, 174, 175*

Iliescu, I., 136-7
immigration
EU involvement at Ukraine-Russia
border, 100-102
EU policy characteristics, 53-4, 55-8
Hungarian policies, 172-6, *174, 175*
see also borders and borderlands;
migration
IMO (International Organization on
Migration), 1102
industries, European
statistics of growth, *92*
interaction, cross-border
Hungarian/Romanian experiences, 166-
7, *166, 167*
see also cooperation, cross-border
International Organization on Migration
(IMO), 1102
INTERREG Community Initiative (EU,
1990), 120, 213
Israel
geopolitical relations with EU, 27-32

Joeniemmi, P., 20
Jordan
geopolitical relations with EU, 27-32

Kassa-Miskolc euroregion, 153, 157
Kohl, H., 17
Kuchmar, L., 96

Laidi, Z., 26
Lamy, P., 24
Law on Local Self-Government (Ukraine, 1997), 88
Law on Local State Administration (Ukraine, 1999), 88
Lebanon
    geopolitical relations with EU, 27-32
Libya
    geopolitical relations with EU, 27-32
Lipponen, P., 23-4
Lithuania
    macroeconomic growth, *92*
Lower Danube Euroregion, *39*, 45-6, 143

Makó Regional Business Zone, 159
management, borderlands
    EU involvement at Ukraine-Russia border, 100-102
markets, labour
    experience in German-Polish borderlands, 209-12
    Hungarian trends and policies, 172-6, *174, 175*
MEDA programme (EU), 120-21
Mediterranean, The
    geopolitical relations with EU, 27-32
Merkel, A., 17
migration
    EU involvement at Ukraine-Russia border, 100-102
    EU policies, 53-4, 55-61
    German-Polish borderland experiences, 209-12
    Ukrainian experiences, 175-6
    see also borders and borderlands; cooperation, cross-border; immigration
Mittelman, J., 33
Moldova
    borderland geopolitics with Ukraine/ Romania, 37-40, *39*
    cross-border cooperation with Romania and Ukraine, 42-4
    cross-border cooperation with Romania, 141-6
    international geopolitical cooperation, 42-4, 47, 48, 133-48, *139*

outcomes of border policies, 37-42, *39*, 44-50
Molotov-Ribbentrop Pact (1940), 135, 136-7
Morocco
    geopolitical relations with EU, 27-32
Mrinska, O., 100

Nastase, A., 137
NATO (North Atlantic Treaty Organization)
    Hungarian involvement, 76-7
ND ('Northern Dimension') strategy
    development and characteristics, 23-7
    value as cross border cooperation perspective, 118
Neumann, I., 26
'New Economy' doctrine, 209-12
New Neighbourhood Initiative (NNI) (EU)
    characteristics, 113-15, 121-3
    involvement of Moldva and Romania, 146-8
    role of 'Northern Dimension', 25, 26
    role in post-Soviet cross-border cooperation, 104
North Atlantic Treaty Organization (NATO)
    Hungarian involvement, 76-7
'Northern Dimension' (ND) strategy
    development and characteristics, 23-7
    value as cross border cooperation perspective, 118

O'Dowd, L., 4-5

Palestine
    geopolitical relations with EU, 27-32
Papandreou, G., 32
partnership
    EU strategies towards Russia, 119
    see also cooperation, cross-border; trade, cross-border; traffic, cross-border
Partnership and Cooperation Agreement (PCA) (EU-Russia, 1992), 24, 119
Patten, C., 32
Peace of Tylza (1807), *197*
PHARE (Poland/Hungary Aid for the Reconstruction of the Economy), 37, 38-9, 89, 120, 146

PHARE-CBC programme, 187, 199-202, *200*
Podilsky region (Ukraine), 85, *84, 86, 87*
Poland
    cross-border cooperation with Germany, 199-204, *201*, 209-12, 213-15
    cross-border cooperation with Ukraine, 186-9
    EU assistance, 187-9, *189*
    geopolitical history, 177-9, 196-8, *197*
    location of euroregions, 202, *203*
    macroeconomic growth, *92*
    migration experiences, 209-12
    structure and administration of Poland-Ukraine border, 179-82, *180, 181*
    see also trade, cross-border; traffic, cross-border
Poland/Hungary Aid for the Reconstruction of the Economy (PHARE), 37, 38-9, 89, 120, 146
Polisky region (Ukraine)
    socio-economic status, 85, *84, 86, 87*
politics, local
    Polish-Ukrainian cooperation, 186-7
politics, national
    Romanian-Moldovan relations, 133-8
Potsdam Treaty (1945), 196, *197*
Pro-Europa Viadrina euroregion, 208-13, *208*
programmes
    funding of cross-border control programmes, 120-21
    German-Polish community, 204
Prydniprovsky region (Ukraine), 83, *84, 86, 87*
Putin, V., 96

regions *see types and names eg* borders and borderlands; Carpathian Basin; Central Region; euroregions
relations, foreign
    Hungarian experience, 74-6
    see also European Union; NATO; trade, foreign; Warsaw Pact alliance
Roman, P., 135
Romania
    attitudes towards cross-border cooperation, 165-9, *166, 167, 168, 169*

borderland geopolitics with Moldova/Ukraine, 37-48, *39*
cross-border cooperation with Moldova, 141-6
cross-border cooperation with Moldova and Ukraine, 42-4
geopolitical characteristics, 41-2
geopolitical relations with Moldova, 47, 48, 133-46, *139*
international geopolitical cooperation, 42-4, 140-41, 143-8, *144-5*
macroeconomic growth, *92*
outcomes of border policies, 37-42, *39*, 47-50
Russia
    cross-border cooperation with Ukraine, 100-107
    EU partnership strategies, 119
    geopolitics of Russia-Ukraine border, 95-100
    impact of EU enlargement, 100-104
    impact of euroregion development, 127-9
    macroeconomic growth, *92*
    relations with 'Northern Dimension', 25-7
    see also regions eg Euregio Karelia

SAPARD (Special Accession Programme for Agriculture and Rural Development), 120
Sassen, S., 58
Schengen regime
    application in relation to borderlands, 40-41
SEA (Single Economic area) agreement (2003), 98
security, borderlands
    EU involvement at Ukraine-Russia border, 100-102
Shlyamin, V., 128-9
Sibley, D., 55
Single Economic area (SEA) agreement (2003), 98
Siret-Prut-Nistru Euroregion, *39*, 143
Slobozhanshchyna euroregion, 107-11
Slovakia
    macroeconomic growth, *92*
Small Projects Fund (SPF) (PHARE), 187

'Southern Dimension' (SD) strategy
    characteristics and geopolitics, 28-32
Southern Region (Ukraine), 85, *84, 86, 87*
Soviet Union
    cross-border cooperation with Ukraine,
        100-107
    EU partnership strategies, 119
    geopolitics of Russia-Ukraine border,
        95-100
    impact of EU enlargement, 100-104
    impact of euroregion development, 127-
        9
    macroeconomic growth, *92*
    relations with 'Northern Dimension',
        25-7
    see also regions eg Euregio Karelia
Special Accession Programme for
    Agriculture and Rural Development
    (SAPARD), 120
SPF (Small Projects Fund) (PHARE), 187
Stability Pact (EU)
    Moldovan-Romanian benefits, 142
Syria
    geopolitical relations with EU, 27-32

Technical Assistance to the Commonwealth
    of Independent States (TACIS),
    37, 38-9, 89, 90, 120, 141, 143-5,
    144-5
trade, cross-border
    Germany-Poland, 202-4
    Poland-Ukraine, 184-6, *185, 186*
    see also entrepreneurship, cross-border
trade, foreign
    Hungarian experience, 70
    Moldovan-Romanian relations, 138-40,
        139
    see also relations, foreign
traffic, cross-border
    Hungarian-Ukrainian, 172, *173*
    Polish-Ukrainian, 182-4, *183*
travel, cross-border
    between Romania and Moldova, 140-41
Treaty on Good Neighbourliness and
    Friendly Relations (1992), 178
Trianon Treaty (1920), 150
Turkey
    geopolitical relations with EU, 27-32
Tunisia

geopolitical relations with EU, 27-32
Ukraine
    borderland conflict with Romania, 44-6
    cross-border cooperation with Romania
        and Moldova, 42-4
    cross-border cooperation with Russia,
        100-107
    geopolitical characteristics and history,
        41-2, 177-9
    geopolitics of Ukraine-Russia border,
        95-100
    geopolitics with Moldova/Romania, 37-
        42, *39*
    impact of EU enlargement, 81-3, 87-93,
        92, 100-104
    outcomes of border policies, 37-42, *39,*
        47-50
    socio-economic status, 83-7, *84, 86, 87,
        92*, 175-7
    structure and administration of border
        with Poland, 179-82, *180, 181,* 186-
        9
    see also trade, cross-border; traffic,
        cross-border
Upper Prut Euroregion, *39*, 143
urban areas
    role in inter-regional cooperation, 159-
        60

Verheugen, G., 22
Versailles Treaty (1919), *197*
Voronin, V., 136

Walzer, W., 56
Warsaw Pact alliance
    geopolitical significance for Hungary,
        68-9, 70-71
    relations with Hungary, 73-4
'Wider Europe' strategy
    commonalities with Russian-led values,
        104-107
    emergence and characteristics, 18-23,
        121-3
    involvement of Modova and Romania,
        146-8
*Wider Europe – Neighbourhood: a New
    Framework* (2003), 21
workers, foreign
    Hungarian policies, 172-6, *174, 175*

Ukrainian experiences, 175-6

Yanukovich, V., 99
Yushchenko, V., 99

Zahony Regional Business Zone, 159
Zimmer, K., 105
Zizek, S., 58
zones, regional business, 159